Meeting
the Expectations
of the Land

ESSAYS IN SUSTAINABLE AGRICULTURE
AND STEWARDSHIP

Edited by
Wes Jackson
Wendell Berry
Bruce Colman

NORTH POINT PRESS *San Francisco 1984*

Gary Snyder's "Good, Wild, Sacred," originally appeared in *CoEvolution Quarterly* (Fall 1983), copyright © 1983 by Gary Snyder; it is reprinted by the kind permission of the author.

Copyright © 1984 by The Land Institute, Inc.
Printed in the United States of America
ISBN: 0-86547-171-1 (cloth) / 0-86547-172-x (paper)
Library of Congress Catalogue Card Number: 84-60686

In memoriam:
    Thomas Jefferson
    Sir Albert Howard
    Aldo Leopold

Contents

Preface and Acknowledgments

A new agriculture is growing in the United States. Born partly of the environmental movement of the 1960s and 1970s, inspired partly by the unending crises to which the American farm is heir, harking back in some senses to the great conservation struggles of the 1930s, paralleling in part the soft energy movement of the 1970s and 1980s, this new agriculture is being conducted by hundreds of people all over the country.

It involves people gardening without chemicals. It involves genetic research on our major grain crops and on certain food animals (the beefalo, for example). It involves developing new ways to own farmland and to preserve it against the land-hunger of sprawling cities, malls, and highways. It involves new attention to the oldest faults of agriculture: its self-destructive tendency to wash away and sterilize the soil, to encourage deserts to develop where people have been raising their food, and to drive the best and the brightest (or, anyway, the most ambitious) people off the land. It involves reexamining the role of food in foreign policy and reconsidering the way we export farming techniques that are questionable in our own Midwest to places in the Third World where they are surely destructive of all that might be valuable in local cultures.

Let us understand from the outset that this new agriculture is not a sub-sistence-oriented, back-to-the-land movement—although that movement has a place here. This change is about the commercial, market-oriented, city-supporting agriculture on which most of us depend. It addresses the fact that this farm system—a wonder of the modern world—does not look to survive much longer. The agriculture that this book discusses has been given many names: sunshine agriculture, renewable agriculture, organic agriculture (a limited term that, like subsistence, has its part in our larger scheme), regenerative agriculture.

What this agriculture features is relatively large numbers of people get-ting their livelihood on the land, growing crops that act like wild eco-systems—that is, that build the health of the soil even as they deliver the seeds (grains), leaves, fruits, meats, and roots that compose a healthy diet. The term we like best for such an agriculture is *sustainable agriculture,* and the definition we like best is Wendell Berry's: "A sustainable agriculture does not deplete soils or people."

This agriculture will start with the places where crops are grown. It will look at the soils, climate, human and natural communities—the whole en-vironment—of a place and then go to work with them to produce food. The soil's needs will be what matters; economics and markets and all the rest will properly meet the expectations of the land or else pass away.

Creating sustainable agriculture will involve agendas for every part of the food enterprise. There will be agendas for farmers, researchers, teach-ers, extension agents, bankers, seed companies, grain companies, makers of implements, suppliers of energy, finally for all who buy and consume food. (In my part of the country, serious eating has become chic, and one of the critical elements of new California cuisine, as it is practiced by Alice Waters, Jeremiah Tower, and others, is the attention chefs are paying to agriculture, to how their food is grown. For example, Alice Waters used to refuse to serve chicken at Chez Panisse because she didn't like the way American chicken was raised; she has since found a supplier. Many of these restaurants maintain their own gardens or have long-term contracts with ranchers, farmers, and orchardists that specify how to raise produce and livestock "cleanly.")

Workers in this revolution's fields will want to remember the Biblical in-junction to dress and keep the earth with crops. This is the rule that the Amish take most seriously, and their survival should tell us something. (The Amish have doubled in number in the past thirty years, rather than being assimilated, as sociologists once predicted.) Their example can help us to meet the expectations of the land.

A word on origins: most of the pieces in *Meeting the Expectations of the Land* were written specifically for this book. Most of them grew out of a weekend of meetings that we held in Des Plaines, Illinois, in June of 1982. Wes Jackson had called us together, at the Holiday Inn there, near O'Hare Airport, to discuss what we all could do for agriculture. Over two days of talk, some of it heated, we hammered out some of the ideas on what such a book would contain. Subjects were shared, thoughts were exchanged, lists of questions to be answered were drawn up. During the book's gestation, such articles as Gary Snyder's "Good, Wild, Sacred," and Donald Worster's "Good Farming and the Public Good," both written for other uses, came to our attention, and we decided to include them. (Worster was at Des Plaines; Snyder was not. Others who joined the project after Des Plaines were Marty Bender, Stephen Gliessman, Hans Jenny, the Lovinses, Gary Paul Nabhan, John Todd, and Angus Wright.) At various stages, the manuscript was shipped back and forth among the contributors, so that they could review each other's work. Thus economists commented on the work of soil scientists, who critiqued the work of ecological researchers and political types.

What I am saying is that this is not an anthology or reader in the usual sense of those terms: a bunch of vaguely related pieces from a variety of sources. Rather it is a gathering of writings that move around a core with which all the writers (although with their differences) agree.

As the traffic director of the editorial crew, I would like to offer the editors' thanks to the following people: to Friends of the Earth Foundation (and especially its president, Alan Gussow, a Des Plaines participant) and Jay Harris (also at Des Plaines) for funding the Des Plaines meeting; to George Young for general moral support; to Robert Schaeffer, of Friends of the Earth, Frances Moore Lappe, and John Vogelsburg for their participation at Des Plaines; and to Linda Okeson, of The Land Institute, and Diana Lorentz for technical assistance. Special thanks go to the students who were at The Land Institute in March of 1983 and who helped Wes and me to thrash out some serious problems: Helen Atthowe, Mark Bohlke, David Burris, Ruskin Gould, Debra Israel, Cary Nailing, Terri Nash, Julie Neander, Alex Stone, and Willow.

My special, special thanks go to Margaret Sheehan for general patience.

Bruce Colman
Salina, Kansas
and Berkeley, California
October 1983

Introduction

Wes Jackson

Anyone can admit that "sustainability" is desirable in agriculture, but the contributors to this volume break from the pack in one important respect—they believe that agricultural problems are too narrowly defined and as a result, so are the solutions. While all of the authors believe that a radical change is necessary if we are to develop a sustainable agriculture, they may disagree about how to achieve sustainability.

Look at the current list of agricultural problems all of us are accustomed to talking about: soil erosion, aquifer mining, the problems and potential problems of pesticides, the energy and dollar cost as well as the health problems associated with commercial fertilizer, the overcapitalization of agriculture, the decline of the rural economy, and the shrinking number of family farms. All are legitimate concerns and the contributors here are deeply bothered by them all. But we see them as symptomatic, as inevitable products of a systemic problem whose roots are in the roots of agriculture, at least as far back as the time thousands of years ago when patches became fields.

A significant part of the problem is that we are still gatherers and hunters, which is another way of saying that we are land animals who were shaped by our evolution to live off the land in highly specific ways. When we conduct agriculture, we are, therefore, altering the ecological arrangement that was responsible for our genesis as a species. I think that this is the reason that this alienation has allowed us to see and regard land mostly as a resource. So we have created a problem for ourselves from the word "go," for land is not a resource any more than humans are resources. Call chrome a resource or petroleum a resource, but not land or people. The concept of resource is restricted to the notion of utility. Land and people transcend a one-dimensional definition that makes economics primary. But when economics is regarded as the brightest star in the constellation of considerations, economic problems are inevitable, for as Thoreau once noted, "the world is more beautiful than it is useful." Should anyone's suggestions for a sustainable agriculture be trusted who doesn't believe that?

The current discussion of the "problem of agriculture" has been much too narrow. We are at that exact instant in history where, as a people, we are discovering another law of nature, this time indirectly, at the point where it affects human culture. Vulgarly stated, the law is that for any level of biological organization—ecosystem, individual, or culture—if a "bottom line" is designated or featured, that feature will break the system. The pattern is clear. It creeps up on us. First the ends justify the means and eventually, to use the phrase of the eminent chemist and critic of science, Professor Irwin Chargaff, the ends "sanctify the means." When the ends merely *justify* the means, there is still time to change, but we are dangerously close to *sanctifying* the means of production agriculture.

Our economic life will remain healthy only so long as other considerations buttress it. Alone it will totter. Economic considerations, taken exclusively, appeal to those unable to tolerate the ambiguities associated with the larger constellation of considerations that impinge on a problem—they satisfy the narrowly analytical mind, the mind given to the sort of thing simple enough to be accommodated by equations and graphs. We have been so intrigued by the *graphs* showing inputs, production, cost, and return that we have felt little need for the *para*graphs. The problem of agriculture may be exposed by economics, but its solution will not.

Solving the problem of agriculture is not on the national agenda as a primary subject for discussion, but it should be. It isn't because the commercialization of food has led to the commercialization of land. Of course, the American natives were more right than the conquerors and settlers. Land cannot be possessed for very long, let alone commercialized. It will eventually claim us, possess us.

These essays attempt to address that problem. If they fail, part of the reason is that our language has not yet evolved to the point where it can accurately describe our proper relationship to the land. That language will only come as we discover the proper relationship of people and land in a modern setting, as we assess not only the tools and techniques but the social, political, economic, and religious arrangements suitable for a highly populated sun-powered planet.

All isn't bleak. The fossil-fuel age has given us a vantage point that a sun-powered culture may never have achieved. What we see from that vantage point is that "sustainability" for both agriculture and culture will not be achieved in a high-energy culture. High energy tends to destroy the elements of sustainability. And now, with the era of shrinking energy upon us, we hope we have broadened the discussion about agriculture. We boast no claim that we have thought of every category which needs to be considered. What we hope to have done, if only through illustration, is to have presented some fresh examples of the kind of thinking and work necessary for a new definition of agriculture, a new synthesis. As disparate as the subjects covered here are, all but two of the essays were written for this volume.

I mentioned at the outset that we don't necessarily agree on the proposed solutions. Three examples come immediately to mind. Marty Strange at the Center for Rural Affairs and Jennie Gerard and Sharon Johnson at the Trust for Public Lands have different perspectives on the role of estate taxes in an idealized future. Hans Jenny, the dean of American soil scientists, believes all organic matter and nutrients should be returned to the soil and that none should be used for fuel, while Amory and Hunter Lovins and Marty Bender advocate deriving some energy from our agricultural lands to fuel the industrial economy.

Two of America's leading essayists and poets, Gary Snyder and Wendell Berry, offer two equally important considerations for resolving the problem of agriculture. Gary Snyder is a student of the paleolithic and early neolithic. He sees humans as the product of a long evolutionary history in a gathering-hunting context, before the few seconds ago, in geological time, that we started agriculture. These patterns of the land are ancient; they are within our bones and nerve fibers and they won't be denied outright. Some of these patterns, in the modern context, are destined to work to our disadvantage, some of them to our advantage. But they are there and should not be ignored, as they have been by countless generations of agricultural policymakers. Wendell Berry, on the other hand, is a defender of culture, particularly the culture which is the product of agriculture, which is in and of agriculture. This cultural tradition was hard-won, just as the patterns of the wild were hard-won for nature. Both are vulnerable in the high-energy industrial world.

Two historians have contributed essays. Angus Wright, a Latin American historian, describes the impact of American attitudes and agricultural science and technology on both agriculture and culture of a Third World country. Donald Worster, an American historian and well-known for his two important books *Nature's Economy* and *Dust Bowl*, has contributed two essays. In one essay, he calls for a "water ethic" as an extension of Aldo Leopold's "land ethic." In another, he describes the terrible predicament of the modern farmer as a problem for all of society, not just the farmer. An agricultural journalist, essayist and small farmer, Gene Logsdon, argues in favor of the importance of traditional farming practices for a sustainable agriculture.

Steve Gliessman defines agroecology and describes how such an approach can contribute to sustainable agriculture. Gary Nabhan, an ethnobotanist who has done important field work, especially among the Papago Indians of the Southwest, explains the possibilities of desert agriculture. Marty Bender and I discuss our investigations into the polyculture of perennials at The Land Institute. Dana Jackson talks about the modest scale at which essentially all individuals can work—the garden, noting that in the aggregate, gardens are a significant source of food. John Todd, codirector and founder of the New Alchemy Institute and Ocean Arks International, describes sustainable agricultural practices in nonEuropean cultures, concluding that truly sustainable agricultural science will be open to influences and data from all sorts of unlikely sources, and will look to local resources for making solutions. My final essay, "A Search for the Unifying Concept for Sustainable Agriculture," attempts to provide an overall perspective and an initial taxonomic framework for understanding varieties of sustainable agriculture.

Table of Metric Equivalents

LINEAR MEASURE

U.S. Unit	Metric Unit
1 inch =	25.4 millimeters 2.54 centimeters
1 foot =	30.48 centimeters 3.048 decimeters 0.3048 meter
1 yard =	0.9144 meter
1 mile =	1609.3 meters 1.6093 kilometers
0.03937 inch	= 1 millimeter
0.3937 inch	= 1 centimeter
3.937 inches	= 1 decimeter
39.37 inches 3.2808 feet 1.0936 yards	= 1 meter
3280.8 feet 1093.6 yards 0.62137 mile	= 1 kilometer

LIQUID MEASURE

U.S. Unit	Metric Unit
1 fluid ounce =	29.573 millimeters
1 quart =	9.4635 deciliters 0.94635 liter
1 gallon =	3.7854 liters
0.033814 fluid ounce	= 1 milliliter
3.3814 fluid ounces	= 1 deciliter
33.814 fluid ounces 1.0567 quarts 0.26417 gallon	= 1 liter

SQUARE MEASURE

U.S. Unit	Metric Unit
1 square inch =	645.16 square millimeters 6.4516 square centimeters
1 square foot =	929.03 square centimeters 9.2903 square decimeters 0.092903 square meter
1 square yard =	0.83613 square meter
1 square mile =	2.5900 square kilometers 1 section 640 acres
0.0015500 square inch	= 1 square millimeter
0.15500 square inch	= 1 square centimeter
15.500 square inches 0.10764 square foot	= 1 square decimeter
1.1960 square yards	= 1 square meter
0.38608 square mile	= 1 square kilometer
1 acre =	4047 square meters 0.405 hectare

CUBIC MEASURE

U.S. Unit	Metric Unit
1 cubic inch =	16.387 cubic centimeters 0.016387 liter
1 cubic foot =	0.028317 cubic meter
1 cubic yard =	0.76455 cubic meter
1 cubic mile =	4.16818 cubic kilometers
0.061023 cubic inch	= 1 cubic centimeter
61.023 cubic inches	= 1 cubic decimeter
35.315 cubic feet 1.3079 cubic yards	= 1 cubic meter
0.23990 cubic mile	= 1 cubic kilometer

DRY MEASURE

U.S. Unit	Metric Unit
1 bushel =	35.239 liters
0.90808 quart 0.11351 peck 0.028378 bushel	= 1 liter

WEIGHTS

U.S. Unit		Metric Unit
1 grain	=	0.064799 gram
1 ounce	=	28.350 grams
1 pound	=	0.45359 kilogram
1 short ton (0.8929 long ton)	=	907.18 kilograms 0.90718 metric ton
1 long ton (1.1200 short tons)	=	1016.0 kilograms 1.0160 metric tons
15.432 grains 0.035274 ounce		= 1 gram
2.2046 pounds		= 1 kilogram
0.98421 long ton 1.1023 short tons		= 1 metric ton

Meeting the Expectations of the Land

1

The Importance of Traditional Farming Practices for a Sustainable Modern Agriculture

Gene Logsdon

During the summer, my chickens get most of their protein supplement by eating blood-gorged flies off the cows. When I let the hens out in the morning, they scurry off to where the two cows lie in the shade, first attacking the big, slow horseflies they can reach from the ground, sometimes jumping up on the reclining cows in their eagerness. When the horsefly population is depleted momentarily, the hens linger around the cows' heads, nabbing the nimbler face flies. The cows never budge during any of these maneuvers, obviously aware not only of the benefits accruing to them but also trusting completely a chicken's ability to peck within an eyelash of their eyeballs without injuring them.

This symbiotic relationship between cow and chicken must be common, or must have been common, on the traditional farm where animals often shared the same quarters. I remember hens on a Minnesota farm roosting on the cows' backs when the temperature sank below zero, but I have never seen any reference before to the cow-horsefly-chicken connection. Having noticed it, I tried to become more consciously aware of the humdrum barnyard activity on my own place, where I attempt to reproduce the traditional farm of about 1940 on a very small (thirty acres) scale. (I use the word *traditional* to refer to those practices of mixed livestock, crop, garden, and orchard farming brought over from Europe and adapted to middle-American climates and soils while being passed down from father to son. There are other kinds of traditional farming, of course, such as that practiced by the southwest Indians.) The first observation I made—one I had hitherto taken for granted—was that the fly population on a cow tended to increase as summer progressed, but only so far, leveling off when there was still more cow to bite and more blood to draw out. What halted the continued population growth? The chickens may have been part of the answer, but obviously only a small part, since they could catch a significant number of only the slower horseflies. And the fly population tends to level off on traditional farms even when chickens do not take a direct interest in the process. This leveling characteristic was certainly not true on the confined dairy where I once milked 100 cows daily. The cows were never let out to pasture there, and the fly population continued to explode until frost, although we fogged the barnlots frequently with insecticides.

The backs of my two cows became a frontier for new discoveries. One day a yellow jacket (ground wasp) buzzed up to a cow and after several unsuccessful attempts, managed to capture a fly and carry it away. I searched books for corroborating evidence but found none. Nor did I see that phenomenon repeated. I yearned for more time just to stand and watch.

For two weeks in August, a flock of cowbirds descended daily upon the cows and sheep. Through binoculars I watched them dart after flies on the cows' backs. They also marched along on foot right beside the cows' noses, not only interested in face flies and other insects stirred up by the grazing animals but also, as far as I could tell, eating bits of the saliva-flecked grass dribbling from the edges of the cows' mouths. A mystery. But I was reminded of a report from the University of Minnesota a few years ago, announcing that a chemical released by the grazing animal's saliva actually seemed to make the grass grow back faster than grass that was cut mechanically. I have seen no follow-up to that report. Knowledge of such ephemeral activity does not program easily in computers, I assume.

I became a ponderer of cow droppings. When the chickens tired of chasing flies, they attacked the cows' manure—what we call cowpies—a normal occupation of chickens and hogs, but one which I never before took the time to observe closely. The hens tore apart the cowpies, reducing their bulk by a third, scattering the fecal matter over a broader area so that the grass was not smothered under it. From the cowpies, seemingly as delectable to chickens as apple pie is to humans, the hens ate partially digested grains and they greedily gobbled chunks of digested grass. The hens were not starved to this diet. They had plenty of fresh grain and grass available to them. They preferred this food. In addition, they pecked at tiny specks of indiscernible stuff and fly eggs, plus other insects drawn to the manure, not to mention earthworms that had worked up from below to dine on the succulent organic matter. Over the course of a summer, I counted seventeen distinct species of insects in and about the pasture cowpies, all working out their life cycles in some kind of tenuous connection with the manure.

This last discovery may have explained another phenomenon drawn into the encircling relationships between farm and nature. The preceding spring, snow lingered into March and I worried that the returning bluebirds would have nothing to eat. During the first thawing days, however, I saw them flying from fencepost to ground and back to post again, just as they do in summer when feeding on insects. I examined the bare ground and meltwater pools between the patches of snow. To my surprise, the scum of the meltwater was dotted with drowned bugs. Where they had come from so early in the year I cannot say, but they resembled closely the gnatlike insects that flocked around the cowpies in the fall, cowpies that in March were miniature islands in the meltwater pools of the pasture.

Awakened to any possibility, I turned my attention to the pair of woodcocks that persists in nesting at the edge of the pasture where trees crowd out from an adjacent woodlot. Woodcocks are fairly rare in this area and terribly shy. Why would they put up with my daily comings and goings and the cows tramping their shady territory to bare earth? In an ornithological report, I found the answer. Cows and woodcocks have their own special relationship. According to the report, cows encourage the presence of woodcocks in two ways: they tramp the ground under trees bare, aiding the woodcocks' search for earthworms, their principal food; the cows' droppings further encourage earthworms to the soil surface to become easy pickings for the probing birds.

The one creature notably missing from the pasture cowpies was the dung beetle. When I was a child, dung beetles—we called them tumble-bugs—were a common diversion. I watched them by the hour as they fashioned their marble-sized balls of manure from the cow and sheep droppings, rolled the balls laboriously down the sheep paths, and then buried them in the pasture grass, each ball with an egg inside it. But about 1950, the dung beetles all disappeared.

Researchers at the University of California and at the U.S. Department of Agriculture (USDA) station at College Station, Texas, have been introducing dung beetles to their respective states. In two days, they say, a pair of dung beetles can bury a cowpie. Since our 200 million cows drop cowpies to temporarily cover an estimated 8 million acres of grass each year, the value of the beetles is enormous—they remove this covering before the grass under it is smothered and bury it before its nutrients are lost by leaching. At the same time, the prompt removal helps control livestock pests, especially face flies and horn flies, which lay eggs in the manure. The latter cost cattlemen $5 million in California alone, say researchers, and pinkeye, spread by a bacteria carried by face flies, causes another $1.5 million worth of damage each year.

In my part of the country, northern Ohio, the face fly appeared in problematical numbers almost exactly with the demise of the dung beetles. But what caused the demise? No one seems to know. It is easy to blame the careless use of insecticides, but more than likely the cause was the abandonment of traditional farming methods. The dung beetle disappeared at the same time that thousands of small-farm sheep flocks were sold and when cows and hogs were taken off pastures in favor of confinement; the permanent pastures were then plowed for corn. The ancient Egyptians made golden images of their dung beetle, the scarab, and placed them in the tombs of their loved ones—a symbol of the return of all that decays to life again. It is a symbol, too, of the basic tenet of traditional farming.

Other interrelationships were yawing off around my cowpies in intersecting orbits. Where the manure decayed in the pasture, white clover sprang up as if by magic. Where the clover grew, the bluegrass followed in a year or two, as surely as night follows day. The grass greedily usurped the nitrogen that clover and rhizobia bacteria, working together, had drawn into the earth from the tons of free nitrogen in the air above each acre. The bluegrass crowded out the clover until the nitrogen was depleted. Then the clover came back to manufacture more nitrogen; the cycle continued as long as grazing and an occasional mowing controlled weeds and brush.

Upon these pastures, the livestock tied together other encircling food webs. They ate from a wide menu: grasses and clovers of various sorts, in addition to the bluegrass and white clover, and more than twenty weed species. They reached up and grabbed mouthfuls of tree leaves occasionally. They pawed and licked the virgin soil of the creek bank, even though they had access to a mineral block in which the experts believe they have put every trace element an animal needs. When certain weeds and grasses went to seed—particularly orchard grass, timothy, Kentucky fescue, and pigweed (amaranth)—the cows and horses took a sudden interest in eating the seedheads, instinctively knowing, I suppose, that these seeds contained more protein than the grains I might otherwise have had to feed them. And what the cows didn't eat, the sheep did. The traditional rule of thumb is that a pasture that will carry twenty cows will carry thirty sheep more and never know the difference. Forty sheep, says sheep- and cattleman Russell Conklin in Kentucky.

All the while, pasturing animals spread their own manure, to keep the whole rich process revolving. John Vogelsburg, on his large traditional farm in Kansas, has carried this idea a step further back in tradition. He has returned to stacking loose hay in his pastures. The cows feed themselves on it in winter, and by moving the location of the stacks each year, Vogelsburg lets the animals spread all their manure instead of having to haul it from the barn himself. He says that with the old stacking equipment, making hay is no harder than with a baler; the old equipment also makes a better quality of hay.

The encircling food webs spinning out from the cowpies entwine the animals finally with the humans who eat them. Traditional farmers put chunks of sod from rich permanent pastures into the pens of sows and new pigs, knowing the animals will derive needed iron from the dirt. The iron in the animal eventually becomes iron in the human. Natural antibiotics live in healthy soil. My dairy calves, pasture born, pasture raised, and mother nursed, have never gotten scours. Confinement cattle, on the other hand, must be pumped full of antibiotics to keep them healthy, resulting in bacteria immune to antibiotics. The enveloping food webs will either produce healthy humans or, if we fail to see the connections, diseased humans. One of our new egg customers called frantically after she cracked the first of our eggs. She thought we would want to know that the egg's yolk was "terribly orangish." She had never before seen a healthy egg.

Not the least significant aspect of the interrelationships between traditional farming and nature is that much of the activity leading to production of food—the purpose of agriculture—proceeds without the expenditure of energy on the part of humans or machines. What machine, however electronically clever, can duplicate the accomplishments of a mere cowpie? Swedish scientist Staffan Delin has recently theorized that "it well may be that the biological processes are many magnitudes of order more efficient than the industrial ones." I suspect that this insight, call it *biological efficiency*, is the key to a practical, sustainable agriculture if mankind is ever to adopt one. For forty years we have tried to apply assembly line efficiencies to farming, coached by simplistic assembly line economics. These efficiencies, it now seems apparent, don't work in farming; they don't even work very well in factories. But to argue that point any longer appears fruitless. A more hopeful course would be to bring civilization's attention to bear on this concept of biological efficiency and find out how it might be used to preserve human culture. The starting point is an intense investigation of what traditional farming has learned by trial and error over centuries of experience, even if that means humbling ourselves to the contemplation of cowpies.

Traditional farms are still in operation, and more farms make use of at least some traditional practices. But the partially traditional farms suffer from a principle basic to any biological system: leave out one strand of the fabric of traditional farming and the whole system falters. One either recognizes all the intertwining webs around the cowpie or one drifts gradually into assembly line economics. This is the wisdom of Amish religion. It may be impossible, for example, to have a sustainable agriculture that does not rotate crops in some way. At least whenever monoculture violates the rule of rotation, agronomic trouble follows.

Traditional farms have several characteristics by which they are known, but above them all hovers a general characteristic in which all traditional practices find their rationale. The traditional farm can survive crisis. It can even survive a series of crises. The urban populations of the Scandinavian countries would have starved to death in World War I and again in World War II were it not for the fabric of Scandinavian rural life: small farms that could go on producing at least sufficient food for the populace even during war. That is why the Scandinavians heavily subsidize their small self-subsistent farms and actually use economic sanctions to penalize their large factory farms. These people know which side their bread is buttered on. Literally.

In 1983, I had a chance to visit at length with a bankrupt farmer, John Nixon, who the year before was farming 3,400 acres of corn, soybeans, and wheat in western Ohio. Partly out of curiosity and partly because misery loves company, he has been studying the reasons for today's rash of farm bankruptcies. Surprisingly (to some), he has found that in every case he has learned about, including his own, the crucial event that broke the farmer's back was not high interest rates, not low prices, not poor management, not high costs, not an overextension of credit—although all these factors contributed greatly to the downfall—but a streak of disastrously bad weather. The traditional farmer would respond, "But of course." Bad weather is going to lay the farmer low sooner or later, and he must be prepared for it. A cash-grain farm, like Mr. Nixon's, is more vulnerable to weather problems than a mixed livestock and grain farm. The greater the variety of crops and livestock, the less vulnerable the farm; the more specialized, the greater the vulnerability. Pasture and perennial crops are less vulnerable than cultivated crops; hail, for instance, does not ruin pasture. As insurance, the traditional farmer has always kept enough extra feed in the barn to last through a bad year. A traditional saying sums up the principle aptly, but modern farmers, following assembly line procedures, have thought they could ignore it: "Don't put all your eggs in one basket."

Traditional farmers keep their eggs out of one basket starting with the way they finance the purchase of land. I once made a study of first- and second-generation farmers in various midwestern communities, using the old county biographical histories. In nearly every case, beginning farmers have had to generate cash from some occupation other than actual farming to pay for their land or to get through lean years. Even where it seemed that the farm was being paid for by farming alone, the farmer was accomplished in some specialized skill that appeared to be part of his farming because he did the work at home. He might have been a sawyer. Or he ran the threshing machine for the neighbors. Or operated a seed cleaner. The farming alone did not pay for the farm. Since the industrial revolution at least, farming has had to operate in an economy geared to manufacturing, with a money growth (interest) tied to factory production capabilities, not to rates of biological growth. A cow never heard of 15 percent interest. Science can push her milk production higher (always at the sacrifice of some other biological attribute) but never as fast as the accelerating exponential interest rates of money, especially when under inflationary pressures.

But the traditional farmer is hardly ever aware that there is a difference between money growth and biological growth. As with all his practices, he either instinctively understands or is taught by a cautious father that to be crisis-proof, don't borrow large sums of money. The single biggest difference between the traditional farmer and the assembly line farmer is that the latter, for better or worse, has had no qualms about taking risks with borrowed money. And by 1982, he was in trouble. Again, traditional farmers have a saying to cover the situation, a saying too often thought of as only a droll bit of folklore from Shakespeare: "Neither a borrower nor a lender be."

Because he won't borrow big money, the traditional farmer is content with a small farm, smaller at least than the "factories in the field." He prefers a small farm anyway, choosing to use biological energy in place of machine energy whenever practical, just as all craftspeople do. Biological energy is limited in quantity and in quality, whether one is making milk or making furniture. But the actual size of the individual traditional farm can therefore vary for the same reason: one farmer, like one cabinetmaker, has more skill, more energy, more desire than another. I once asked an Amish farmer who had only twenty-six acres why he didn't acquire a bit more land. He looked around at his ten fine cows, his sons hoeing the corn with him, his spring water running continuously by gravity through house and barn, his few fat hogs, his sturdy buildings, his good wife heaping the table

with food, his fine flock of hens, his plot of tobacco and acre of strawberries, his handmade hickory chairs (which he sold for all the extra cash
he really needed), and he said: "Well, I'm just not smart enough to farm
any more than this *well*." I have a hunch no one could.

A second characteristic of the traditional farmer, implicit in the one just
mentioned, is his or her acquisition of many farm-related skills. He can
build barns and houses and knows how to grow the wood to build them
with. He is a fair veterinarian, an expert mechanic and welder, can wire,
paint, and plumb a house, lay cement, ditch a field, butcher a hog, and fix
almost anything with baling wire and a pair of pliers. My neighbor Bob
Frey built his own tractor with tires from a junker, two transmissions from
wrecked cars, a motor from a burned-out truck, and a frame he welded
together himself. It has twenty-two forward gears and has been running
for twenty-seven years.

A third characteristic, implicit in the second, is that the traditional farm
is a place of varied enterprises. To spread his labor effectively over the
whole year, the farmer has many, but relatively small, sources of income
rather than one or two large sources. Elmer Lapp, who farms in Pennsylvania, sells horses, cows, milk, hogs, honey, eggs, guineas, pigeons, chickens, fruit, ice cream, flowers, collie puppies, cats, and tours of his farm.
There are even fish in his horses' water tank. With this variety of enterprises, Lapp's work involves a marvelous synergy. When he is hauling manure, he is not only fertilizing his field and cleaning out the barn but also
saving the cost of purchased fertilizer, adding organic matter to the soil,
controlling erosion, exercising the horses, training a new colt, making the
cows comfortable and their milk purer, and keeping his eye on the life of
his farm. I have noticed that farmers who do not have manure to haul in
slack times tend to spend the hours in cafes complaining about how poor
farming is these days. With varied enterprises, the traditional farmer has
something to sell every month. Tradition long ago solved for him a concept
the computer boys think they discovered: cash flow.

But none of the enterprises is so large that extra money must be spent
keeping track of them. The traditional farmer does not need to employ an
accountant. He rarely must go outside the family to hire help. His herd of
cows is not so large that he needs a computer to figure his feed rations or an
electronic eye sensor in the feed bunk to dole the rations out properly. One
dairyman I recently visited told me he didn't even believe in the expense of
keeping Dairy Herd Improvement Association records. "I *know* which of
my thirty cows is giving enough milk and which ones ain't," he said. Nor
does the traditional farmer ever have to put big money into new, large

models of machinery. Smaller, older ones will usually work in his small fields. Horses, for many jobs, are even better.

Because he deals in smaller amounts of many different things, the traditional farmer usually has time to let his grain dry naturally, avoiding the considerable expense of artificial drying. Field-dried small grains and crib-dried ears of corn are generally of better quality and eliminate the risk of purchasing grain overheated by an elevator drier or ruined in a moldy pocket in an elevator tower; commercial feeds also are often full of weed seeds. Because he feeds his grain mostly to his animals, he has no large trucking costs, nor is he paying the elevator's storage fee or handling fee for purchased grain. He avoids expensive supplements that supposedly speed the fattening process or the milk flow. He has found his own good grain and hay to be sufficient. He has only so many hogs, sheep, and cattle going to market every year anyway, and it is not important that they all fatten up quickly and are ready to sell at the same time. He has learned that selling a portion every month works out to an average price generally as good as trying to hit the high markets some expert wants to tell him about—for a fee.

In the field, he is just as financially conservative. He follows a traditional rotation alternating row crops with hay and pasture crops, applying manure on the old hay sod ahead of corn, following corn with soybeans or small grains like wheat, oats, or barley, and interplanting clovers and grasses in these small grains to come on for hay and pasture the next year. These rotations enable him to save about half of the fertilizer bill and almost all of the assembly line farmer's herbicide bill; they also solve many insect, disease, and soil compaction problems.

My neighbor with the homemade tractor follows the old economy of grazing animals all over the farm at appropriate times, so that in addition to all the other synergistic activities taking place in rotated crops, the grain crops themselves also double as temporary pastures. He takes the sheep off the permanent pasture in July and puts them in the wheat stubble after wheat harvest, where they nibble the new clover growth, clean up weeds, clean out the fence rows, and glean any grain left by the combine. Then he moves them to the oats stubble right after harvest where they do the same as in the wheat stubble. Then the fattening lambs go into the standing green corn in August, where they eat the lower corn leaves (making corn harvest an easier job), eat weeds grown up since the last cultivation, and clean out the fence rows. They reach up and eat only a few ears—the only grain they will eat before going to market. Later these lambs go onto the old hayfield that will be plowed for corn and graze there until November.

A catastrophe? Traditional farming has an antidote even for tornadoes. Two weeks later—*two weeks*—the Amish community, donating its labor, had sawed and milled the downed timber into lumber and from it built a magnificent new barn. I watched dumbfounded as the barn went up in two days. Carpenters estimated the cost of the barn to a conventional farmer at over $100,000. The Amish farmer told me it cost him $30,000 in actual cash, most of which he received through the Amish church's own insurance program. The barn can hold more than a year's supply of hay, straw, and grain for the animals it houses in comfortable stalls and pens below: twelve to fifteen cows and their offspring, a bull, eight to twelve sows and their pigs finished to market weight, and eight horses. All without one watt to make it functional. What assembly line economics can match this community action? Why is it "old-fashioned" for a community of farmers to help each other out this way? Their labor is the one commodity they can share that no power can take from them.

The draft horse provides a good example of the traditional grasp of biological efficiency as opposed to assembly line economics. By the criteria of assembly line economics, work horses seem to cost more than tractors and the arguments pro and con have continued for fifty years. True Morse, then president of Doane Agricultural Service, Inc., made one of the first explicit statements of the assembly line viewpoint in 1946 in *Farm Journal* magazine. He pointed out that while it did cost more in actual cash outlay to keep a tractor than a team of horses (even then), the farmer would be ahead if he sold the horses and bought two cows in their place. Farmers took that advice, sure enough, and we have had a surplus of milk ever since, along with an army of bureaucrats to administer a tax-supported market subsidy program to direct an agenda that USDA economists have admitted to me is "too complicated to explain." But that point aside, the divergence of view between biological efficiency and factory economics as it pertains to horses stems from how each side defines cost. Morse put the annual cost of keeping a team in 1946 at $182.49. (The traditional mind immediately suspects such finely wrought numbers. Why not $182.50? A computer might be forgiven such preciousness, but not anyone who has smelled horseflesh. I find myself wondering if Mr. Morse figured in the wear and tear on the harness, or the fact that old Bell hangs back in the traces worse than Flora, or if he counted the penny of work the dung beetles contributed to hauling off the manure.) But even in the unlikelihood that horses can be summed up by numbers so precisely, what does this cost mean? To a farmer who loves horses, a good team bears satisfaction akin to the modern playboy's love for his Corvette. It is a cost that might easily be reckoned as a profit, for horses are worth much more to the horse lover than the money he might save milking two more cows the rest of his life.

More importantly, most of Mr. Morse's cost is not out-of-pocket cost, a distinction an accountant is bound to scoff at but not a farmer who knows, especially in the 1980s, what it means not to have any cash in his pocket. When that happens to a tractor owner, he still must find money to pay the tractor bills. He has to borrow it, if he is "credit worthy" as the bankers say so unctuously. (If he is "credit worthy," he has the honor of having to pay interest that has reached 15 percent in recent times.) A horse doesn't charge a farmer interest. The bartering system between the horse and farmer is immune to inflation and depression. The horse trades its labor, manure, and affection for food and shelter, then throws in a free colt every year or so that is generally worth more than any of the farmer's calves. No cash outlay is involved at all except for veterinarian bills. The horse will start every morning no matter how cold the weather, and it will run all day no matter what happens in the oil fields of Saudi Arabia. That is the kind of security that biological efficiency provides. Beats bankruptcy any day.

Mr. Morse provides another piece of assembly line logic. The cost of the horses per hour, in 1946, was determined to be about 13¢; the cost of the tractor per hour was about 47¢, over three times greater. But, Mr. Morse reminds us, the tractor did the work of eight horses (that's arguable—the 1946 tractors I've driven will hardly pull a loaded manure spreader in winter snow and are almost useless for dragging logs out of the woods). But assuming Mr. Morse is correct, what he did not add was that the tractor took the place of three or four farmers too. When farmers chose tractors, 75 percent of them chose their own demise, their land becoming someone's factory in the field farmed at three times the cash outlay they had spent farming it.

How much money would farmers have saved not buying tractors between 1946 and 1982? Tractors rose in price from $800 to $18,000 for comparable models. The horses kept having colts. Twenty horse farmers, for example, each with 100 acres, would have had no cash outlays for new models. The modern tractor farmer, going broke on the 2,000 acres those twenty farmers would be farming, owns (with the bank) around $200,000 worth of tractors. In 1955, another farmer, Raymond Rall, leaned over his tractor fender and said to me indignantly: "Farming with tractors, I'm spending up all the money I saved farming with horses." Ironically, yet another neighbor, Jerome Frey, who has kept his horses, just for fun, confessed the other day that he believed he could make just as much money farming 80 acres with horses as on five times that much land with tractors. "But, heck," he shrugged, "I'm kind of stuck with all that land now."

The net profit from the two cows to replace the team of horses in 1946 has not kept pace with the surging tractor prices either. Mr. Morse quotes a cash return per cow in 1946 of $119.50 per year. According to *Dairy-*

man's Digest, a cow in 1982 giving 16,000 pounds of 3.5 milk netted $106. This figure deducts $318 per cow for labor, a deduction Mr. Morse probably did not cipher into his cash return in 1946, but nevertheless, cows have hardly proved to be that much better an investment than horses. Using assembly line economics, one must now keep 100 cows, if only to pay the labor it takes to milk 100 cows.

The traditional mentality approaches profits differently, substituting biological quality for assembly line quantity. In Ohio, Rex and Glenn Spray have bred their Holsteins for high butterfat content, not high production, for ten years—in direct opposition to the advice of assembly line economists. Their thirty-five cows today average 4.1 percent butterfat. With a premium of 17.1¢ per point for every decimal point above 3.5, their six extra points amount to an extra dollar per hundredweight for their milk, or an extra $160 from a cow giving 16,000 pounds of 4.1 percent milk. On thirty-five cows, that is a $5,600 annual premium at no out-of-pocket cost at all. With the current figure of $106 net profit per cow, to make that $5,600 from production alone requires another fifty cows.

Once the minds of farmers and scientists open to the possibilities inherent in biological efficiency, what wonders will follow? In an example so outlandish one suspects it is a joke, *Audubon* magazine reports that a greenhouse grower in Oregon is heating his greenhouse with rabbits. I assume he puts the manure to good use, too. Experiments at the Rodale Research Farm in Pennsylvania and at the University of Southern Illinois indicate the feasibility of an old Oriental tradition: fish can be fattened directly on the manure of chickens and hogs, and the pond water then makes an excellent fertilizer. The Chinese get amazing protein production combining fish and ducks in the same pond-rearing system.

As traditional American farmers pursue the "let nature do it" methodology, they may find, as I have, that they edge closer each year to a type of farming based upon permanent or semipermanent pasture arrangements and lesser amounts of annually cultivated land. The key to general success in this arrangement is a change in marketing standards. Like myself, Oren Long in Kansas raises a Jersey-Angus cross of baby beef (I have Guernsey-Angus cross) that when butchered at 700 pounds makes a most delicious meat, fattened on ample milk and pasture but no grain, the mothers getting only a little grain during lactation, or none when pastures are lush. (In addition, we get our own milk and cream from the cows while the calves are fattening, for greater biological efficiency.) But this meat will not be graded as "choice" because it is not grain fed. Therefore, the farmer cannot sell it profitably in the conventional market. He must develop his own customers who know good meat when they taste it. In the same way, our lamb is luscious even to people who previously said they "hated mutton" and we have produced it quite successfully on mother's milk and weedy rough land, without even good pasture, let alone grain, involved.

Small-bred cows, like Jerseys and Guernseys, will produce a fair to good amount of high-quality milk on good legume hay and very little grain. Heavens, if the public wants low-fat milk, as the milk marketers say, add water to it after it leaves the farm. Why make the cow do it? Then those who appreciate good rich cream and creamy milk can have their desires met in the marketplace, too. More and more of the milk and especially cream in stores now is "ultra-high temperature" pasteurized stuff that will keep without refrigeration but bears a bitter aftertaste (not to be confused with the off-taste from milk kept in plastic jugs). The dairy industry wonders why milk consumption is down. The crowning irony is that I cannot sell my own milk, from cows checked annually for tuberculosis and brucellosis as in any Grade A dairy, cows who have never had mastitis and never been treated with antibiotics. It is against the law in Ohio to sell raw milk no matter how good and healthful it is. The law says I can drink it, but my neighbor can't, which reveals the true purpose of the law: to stop me from underselling the assembly line milk production system, which I can do because of the biological efficiency of a very small operation.

Even nongrazing animals make excellent use of semipermanent pastures. Hogs can fatten profitably, says Morrison, with up to one-third of their diet derived from alfalfa pasture. A small flock of hens can find half its food or more ranging the barnyard and adjacent fields. I have halved the amount of grain even broilers need, ranging them outside this way. The only difference is that they take a little longer to reach butchering weight.

If marketing standards were changed, a successful farm could be maintained at a ratio of 10 acres in semipermanent and permanent pastures to 2 acres of annually cultivated crops, the whole not exceeding about 200 acres and probably working best with about 120 acres of which 20 were woodland for lumber and fuelwood. At this size, biological energy could be substituted profitably for the large inputs of mechanical energy necessary on the assembly line farm. A great variety of products could be sold: milk, cream, lambs, wool, beef, eggs, fuelwood, lumber, honey, fruit, broilers, hogs, surplus horses, old cows, and hens. Of the 120 acres, only 20 would be in annual cultivation, and that 20 would be divided into three fields rotated from corn to wheat or oats to legume hay. The cultivation of these three small fields could easily be handled with horse power or very small tractors. The other 100 acres in permanent and semipermanent pastures would need to be fenced into relatively small plots, the animals moved from one to another regularly after the New Zealand method. Hay would be harvested from about half the pasture whenever the grass and legumes grew faster than the animals could graze them. Haying would be the farm's heaviest labor and fencing its highest cost. Vogelsburg's haystacks and New Zealand-type electric fences are two answers to lessening these burdens. More will come.

But providing the technical information from tradition that would help forge a modern sustainable agriculture is not the most difficult task. Much of the technology is there waiting to be improved. What we lack most is the traditional farmer and a culture that could nourish him. The human being should not be forced into a servile form of farming. Farmers must derive happiness and humane satisfaction from a life that eschews the kind of consumerism, leisure, and delirious pursuit of novelty that characterize our society. There can be plenty of healthy leisure and novelty on the traditional farm, but it must be taught or discovered just as the technology of traditional farms must be taught or discovered. The reason the latest so-called back-to-the-land movement failed for so many people is not so much that they were ignorant of traditional technology—too stupid to stuff rags in a rat hole, as country people say—but that their homesteads were islands in an alien culture. There was no community to rebuild their barns or their dreams. Indeed, their retired parents were often hostile to such "outlandish" dreams of forsaking the American way of luxurious consumption, and their teenage children rebelled against physical labor for which there was no cultural reward or recognition.

To husk corn alone, in a cold December field, day after day, is a misery, one that the traditional farmer was driven to, unconsciously, by the influence of assembly line economics on his way of life. No human is going to accept that kind of work if there is an alternative. But before the industrial revolution, corn was hauled in good weather to the barn and the young people went from farm to farm through the winter evenings making a party out of the husking. The person who husked a red ear—and there were many red ears in the days before standardized hybrid corn—got to kiss his or her sweetheart. In remaking an agriculture that is technically correct for sustainability, we must make sure it is also culturally correct or it will not succeed.

2 | Whose Head Is the Farmer Using? Whose Head Is Using the Farmer?

Wendell Berry

On April 12, 1982, *Newsweek* published in its "Business" section a worried article entitled "Down and Out on the Farm," which made much of "the worst economic conditions to hit the Farm Belt since the 1930s." Recession, high interest rates, high costs, bad markets, bad weather, etc., were (again) driving farmers out of business. Export sales, which supported "the booming farm industry of the 1970s," were expected to decline in 1982. "Farmers' debts," according to the article, "now total $194.5 billion, more than twice the amount in 1976—and today each $1 of farm income must support debt of more than $12, versus a $1-to-$2 ration as recently as 1973."

In the spring of 1982, we were struggling with the fact that another bumper grain crop would cause another round of farm bankruptcies and other hand-wringing articles in magazines such as *Newsweek*. For nothing at present is more destructive of farms and farmers than bumper crops. High production keeps production costs high and the market prices low— a bonanza (so far) for the industrial suppliers of "purchased inputs" to farmers and (so far) to the banks from which desperate farmers must borrow (at usurious interest) the increasing amounts of money necessary to bridge the gap between a depressed farm economy and an inflated industrial economy. But high production is death to farmers. Though most interested parties seem still addicted to the perception of American agricultural productivity as a "miracle," even news articles now occasionally glimpse the fact that, for farmers, a good year in corn is a bad year in dollars.

There was an unacknowledged gravity in *Newsweek*'s story of an Indiana farmer (a grain farmer, one supposes) who "has been forced to put the land he owns up for sale to pay off his $400,000 debt—and hopes he'll have enough left to begin a dairy farm, one of the few areas of farming doing well these days." For dairies are "doing well these days" because of price supports *without production controls*, with the inevitable result that there is now a huge government-owned surplus of dairy products and growing political pressure to lower the supports or end them altogether. The last thing dairy farming needs is an influx of failed grain farmers to further increase production.

And yet farmers and farm experts alike cling to the notion that high production is a "solution." According to the *Washington Post* of June 13, 1982, "about a fourth" of the U.S. Department of Agriculture's $430 million research budget "is aimed at improving farm productivity," whereas "less than $1 million" is devoted to studies of organic farming methods that would reduce the farmer's production *costs*. Farmers who lost money growing corn in 1981 tried to bail themselves out by growing as much or more corn in 1982. The standard advice to dairy farmers is still likely to be that they should voluntarily reduce production—by culling their least productive cows. And in the midst of the overproduction worries of 1982, I heard an expert on the radio telling farmers that they could make it through these hard times by "high production and smart marketing"— apparently without the least flicker of a doubt that high production could get them out of the same jam that high production got them into or the least glimmer of a realization that some markets drastically limit the efficacy, and even the possibility, of "smart marketing."

A number of things about the *Newsweek* article are at once characteristic of most agricultural reporting and symptomatic of the real (largely unreported) difficulty that farmers and farming are in:

1. It is written by journalists who are by no means farmers, who observe agriculture only distantly and from the outside, who notice it at all only when some calamity makes it "news"—and who therefore know little about it.

2. By a rule of journalism, an article about an agricultural depression must exhibit the sufferings of several individual farmers. This is the doc-

trine of "human interest," which holds that we must remember that public events at some point affect the lives of "real people"—which, at first glance, appears to be a decent idea. The trouble is that there is a considerable difference between "real people" and human exhibits produced to illustrate a set of figures. The *Newsweek* writers think that human misery is somehow—mysteriously and surprisingly—*added to* the economy. They don't see that the economy itself is based on a number of miserable assumptions and that it produces exactly the human suffering that is implicit in it. These writers cannot tell us much about agriculture because they do not know much about it; they do not know much about it because from the distance of journalistic observation they cannot penetrate the economy's fraudulent equation between agriculture and industry, farming and business.

3. At the end of their article they trot out (a little doubtfully this time, to their credit) the mandatory expert who welcomes the economic crisis as a necessary means of weeding out the inefficient farmers: "'I think the whole thing is going to make the farmer a more sophisticated businessman,' says Will Schakel, director of the commodity department for the Indiana Farm Bureau."

Although these writers know that the agricultural economy is in trouble, causing trouble to people, and they have the figures to prove it, they have no notion why it is in trouble. This article does not disturb at all the general acceptance of the industrial ideal under which farmers have been laboring, and failing, for half a century or more: the highest possible annual production by the smallest possible number of "workers" for the least possible monetary return.

What the *Newsweek* writers do not see is that the farm "crisis" of recent years is only an episode in a history of agricultural loss—of topsoil, farmers, farms, and farmland—that has been critical, and steadily worsening, ever since the Second World War, a history not noticeably abated by "the booming farm industry of the 1970s." They do not see, either, how agriculture fits among the other needs and obligations of human life. And so they find it easy to write as if agriculture were only a matter of economics, of producing and marketing, which has, incidentally, the power to affect the lives of the dwindling number of farmers. They do not go so far as to comprehend that the influence goes the other way as well—that the lives of farmers affect agriculture and that agriculture can affect the economy. There is no acknowledgment in their article that agriculture uses land and produces food and that, therefore, the related losses of soil, farmland, and farmers raise troubling questions about the continuing availability of enough to eat. They conclude with a vague sense of urgency—"the farmers who survive the current blood bath will be candidates for the agricultural hall of fame"—but nowhere do they betray any premonition that the survival of farmers might involve the survival of anyone else.

This *Newsweek* article is for the "general public." Agricultural journalism of the conventional sort, so far as I can see, is simply the other side of the coin: the same vague, disconnected agricultural gabble, but having, however, a much clearer purpose, which is to sell agricultural machines and chemicals to farmers.

I listen fairly often, for example, to a radio "farm show" that comes on in the early weekday mornings to catch farmers in the dairy barns or at the breakfast table. This program is a mixture of country-flavored music, weather reports, agri-industrial commercials, advice from university and USDA Extension Service experts, and interviews with successful farmers. The tone is set, unsurprisingly, by the commercials. No free solutions are recommended, either by the sponsors or by the experts. The successful farmers are successful by standards unembarrassing to sponsors and experts. The experts, ex-farm boys proud of their success in escaping the farm but uneasy about it in the presence of farmers, are chummily condescending: "Boys, let's get the spray on those alfalfa fields."

And not only must all recommended solutions be purchased, but only infallible solutions are recommended. It is never proposed that a recommended solution merely *might* work. It is never hinted that, under any circumstances, a recommended solution might be too expensive or unnecessary. Since all needed solutions are readily available—at a price, though the price is never mentioned—the message is that farming is a business in which there are no real problems. The voices are *supremely* confident.

So is the voice of the weatherman (now called a "meteorologist") who comes on at intervals to tell us exactly what to expect from the weather (the same weather known to some farm experts, in moments of clairvoyance, as "a variable"). This meteorologist is frequently, embarrassingly, wrong. On some supposed-to-be clear days it rains before ten o'clock. Sometimes it seems that this meteorologist could improve his accuracy by turning away from his "action track color radar" and looking out the window. One could forgive these failures—as one could maybe even forgive the failures of recommended costly solutions; to be wrong, after all, is human—if these voices only sounded human. They don't. The meteorologist's voice is precise, expertish, imperturbably confident, implying, not that he has forgotten yesterday's mistake, or that yesterday's mistake was a rare fluke, but that he was right yesterday, is right today, will be right tomorrow and forever.

These people undoubtedly have lives, at-home lives, that have certain things in common with the lives of farmers: uncertainty, worry, bad surprises, loss of confidence, fear of failure, failure. But the world of radio agriculture, from which they speak to farmers, is not the same world that farmers live in. Even though this is a "farm program," it comes from a world that, for one thing, is persistently urban. These people think that daylight-saving's time saves daylight. They think that there is such a thing as a "weekend" on which, by inalienable right, humans may be free of responsibility for forty-eight hours. They work indoors; sunlight is a phenomenon that they hope to bathe or play golf in on Saturday and Sunday. One morning when we were in a dry spell that was fairly serious for farmers, I heard the meteorologist announce an approaching storm system that might bring us half an inch of rain—"the only bad part" being that it might continue into the weekend.

But the world of radio agriculture is also impermeably general, simple, and dependable. It is a world in which the weather is predictable, in which "purchased inputs" invariably mean high production and high production invariably means high profits, in which pesticides poison pests without fail and poison nothing but pests. It is a world, in short, in which people live in their minds—or are privileged to seem to do so.

This journalistic-commercial version of agriculture may be contemptible, but it is nonetheless powerful. It says that farmers and consumers alike dance to a tune called by the industrial economy; it says that this is both a good tune and the only tune. To both farmer and consumer it says that private knowledge, judgment, and effort can be satisfactorily replaced by generalized, expensive technological solutions. These solutions are never recommended in terms of a precise understanding of specific local needs or conditions, but are offered to the country and the public at large and with the support (so far) of scientific authority readily available, not only in the industrial corporations but also in the governments and the universities.

This industrial version of agriculture is paid for—and this is my point—not by money only but by the intelligence of the buyers as well. The products offered for sale by the makers of agri-industrial technology (like the products of the food-processing industry) are not just ready-made solutions; they are ready-made thoughts. And the larger the scale of the technology and the thinking, the harder it is to modify these thoughts by the action of individual intelligence. It is well understood by now that there is a limit of technological scale, of farm size, of expenditure and indebtedness beyond which the farmer enters a kind of mental paralysis, having simply no choice but to continue to do as he has done until the bank closes him out. He cannot change because he has no margins to turn around in. At this point he is literally dancing to a tune called by an economy that has always (as he sees too late) proposed his failure as the price of his participation.

One way of describing the action of the industrial economy is to say that it comes between the mind and its work. If a mind is to be *on* its work in such a way as to do it well and to preserve in all respects the possibility of good work in the future, then obviously the amount of the work must be limited. Given the right amount of work, the mind lives in its place, not merely as owner or user, but as a fellow creature with the other creatures that belong there, the effective husbander of both the agricultural and the natural households. A mind overloaded with work, which in agriculture usually means too much acreage, covers the place like a stretched membrane—too short in some places, broken by strain in others, too thin everywhere. The overloaded mind tries to solve its problems by oversimplifying itself and its place—that is, by industrialization. It ceases to work at the necessary likenesses between the processes of farming and the processes of nature and begins to order the farm on the assumption that it should and can be like a factory. It gives up diversity for monoculture. It gives up the complex strategies of independence (the use of manure, of crop rotations, of solar and animal power, etc.) for a simple dependence on industrial suppliers (and on credit).

Once the mind is divided from its work—once qualitative problems have begun to be "solved" by quantitative solutions—then no amount of additional mechanization, automation, remote control, computerization, scientific research, expert advice, or stress management, can stop the damage, much less heal it. All those would-be solutions are based on a mistake about the kind of work good farming involves and about the kind and quality of mind required to do that work. This mistake is now not only firmly established in the government and universities; it is widely believed by farmers themselves. A good farmer I know said recently, referring to this mistake: "The farmer today doesn't use his own head." And speaking of the willingness of many farmers to buy caps bearing the names or emblems of "agribusiness" corporations, he added, "He uses his head to advertise products."

The mistake is in the assumption—centuries old, but more damaging in our time than ever before—that the agricultural mind is an extremely *simple* mind, dulled by bodily labor: the mind of a "hick" or "yokel." This assumption exactly suits the purposes of the agricultural industrialists. The last head they want the farmer to use is his own. They want to do his thinking for him, because it is enormously profitable to them to do so. They don't, of course, tell him that he is a stupid yokel; they tell him that his problems are simple and can be simply solved by buying their expensive products.

The industrial version of agriculture has it that farming brings the farmer annually, over and over again, to the same series of problems, to each one of which there is always the same generalized solution, and, therefore, that industry's solution can be simply and safely substituted for his solution. But that is false. On a good farm, because of weather and other so-called variables, neither the annual series of problems nor any of the problems individually is ever quite the same two years running. The good farmer (like the artist, the quarterback, the statesman) must be master of many possible solutions, one of which he must choose under pressure and apply with skill in the right place at the right time. This solving requires knowledge, skill, intelligence, experience, and imagination of an order eminently respectable. It seems probable (in farming as in art) that such a mind will work best which is informed by a live tradition.

What kind of mind does a good farmer have? It is probably not a mind that can best be measured by its ability to retain information or solve isolated problems. We can begin to answer the question, I think, only by trying to suggest something of the nature of the structures of good farms.

In my own part of the country, the best traditional farming is grass farming. This is necessary because of the rolling, sometimes steep topography and clay soils that erode easily. The annual cropping pattern on a 200-acre farm would be more or less as follows. In the spring, ten acres of sod would be broken, three acres for tobacco, seven for corn. The preceding year's row-crop acreage would be in small grains: the tobacco ground planted in wheat in the fall, the corn ground planted in wheat or rye in the fall or (after a late harvest) in oats in the spring. On frozen mornings in late winter or early spring, these wheat or rye or oat fields would be seeded with red clover and grass. Thus in any year, less than 10 percent of the farm would be planted to crops that require disturbance and exposure of the soil. In years when all of the row-cropped land could be sowed in small grains in the fall, only ten acres (5 percent) would be disturbed. Each year would add ten acres of red clover and grass, harvested for hay for two years, and then returned to pasture. Such farms—and there were *many* of them until about a generation ago—would be stocked with cattle and hogs, and often with sheep as well. And so the cropping rotation, shifting around the farm because of the requirement that only sod ground should go to row crops, must mesh with a grazing rotation.

In the more level land of the corn belt, a larger acreage is given to row crops and small grains, but still, on the best traditional farms, the cropping pattern is varied and complex. I will quote from my own description (*The*

Gift of Good Land, p. 252) of a good Amish farm of eighty-three acres in northern Indiana. The farmer here told me that "every fall he puts in a new seeding of alfalfa with his wheat; every spring he plows down an old stand of alfalfa, 'no matter how good it is.' From alfalfa he goes to corn for two years, planting 30 acres, 25 for ear corn and five for silage. After the second year of corn, he sows oats in the spring, wheat and alfalfa in the fall. In the fourth year the wheat is harvested; the alfalfa then comes on and remains through the fifth and sixth years. Two cuttings of alfalfa are taken each year." Here too the cropping pattern must mesh with a livestock pattern, involving, on this farm, Belgian horses, Holstein cows and heifers, and hogs.

I assume that even people without any farming experience can see that these structures, as I have so far described them, are already too complex to be simple-mindedly managed by "yokels." The rotations must be kept revolving smoothly and without interruption in spite of the inevitable disappointments, surprises, "variables," and other annual differences. And for this to continue year after year without damage or diminishment, certain balances must be carefully kept: there must be a balance, for instance, between crops and livestock, with respect both to nutrition and to fertility; there must be a balance between grain crops and legumes; there must be a balance between work and worker.

But so far I have described only the most fundamental structure of a farm, its revolving floor-plan, what we might call its pattern of use. On a good farm that pattern must work with and within a pattern of maintenance. Fences, buildings, and equipment must be protected, repaired, and replaced as necessary. The fertility of the soil must be maintained or improved; this is done partly by the patterns of cropping and grazing, but it also involves questions about where and when manure will be put on the fields, about the use of commercial chemicals (if any), about various strategies of erosion control. And the health of the livestock must be maintained —first of all by and within the general health of the farm (healthy crops make healthy animals), but also by proper attention to brood stock, to the problems of conception and birth, to diseases and wounds.

The involvement of livestock in farming (and I think it has almost always been involved in the best farming) greatly complicates the work—and the mind—of the farmer. The feeding of livestock is, by itself, an intricate science and art. Not only must the different species of stock be fed differently, but each species must also be differently fed for different purposes. One feeds replacement heifers, for instance, differently from the way one feeds brood cows, which are fed differently from slaughter steers. And to be a good feeder it is necessary to be an able judge of the quality of feed, and of the quality and condition of stock. One must pay attention—not just throw the feed in, but stay and watch; see if it is eaten with relish and cleaned up; see if all the animals are eating, if all are healthy, if the feed is doing the good it is supposed to do. It is obvious that if such work is to be done well, it cannot be done mechanically or automatically; an able intelligence has to be involved.

But to me nothing so forcibly suggests the competence and the fineness of agricultural intelligence as the subject of livestock breeding. I mean, of course, *selective* breeding, for it is in selection that intelligence is applied. This process of selection goes back thousands of years. Long before stud books or records of any kind, long before there was a science of genetics, farmers were engaged in the slow sculpture of breed differentiation and improvement. It could be argued that nearly all the *great* work of agricultural science was done before agriculture became a science and perhaps for that reason. There have always, undoubtedly, been great stock breeders, stock-breeding geniuses. These have been people who could carry an ideal or imagined pattern in mind, perhaps all their lives, and work toward it, exactly as a sculptor works to realize an imagined pattern in stone or bronze. They are remarkable, among other things, for their ability to recognize "their kind" in the absence of live examples—to assemble, say, a band of draft mares or a flock of ewes, precisely uniform in conformation, by picking up one at a time, if necessary, at farms and sales far from home.

But just as important as the need to recognize the preeminence of such individuals is the need to recognize that they have never been solo performers. Their work has depended on the work of many others, good farmers and good stockmen, whose intelligence, judgment, sensitivity to local needs and conditions, and devotion to high quality were at once excellent and ordinary. The idea of ordinary excellence seems paradoxical, perhaps simply impossible, to the modern mind, but it is one of the common necessities of a healthy agriculture and a healthy culture.

According to the British farmer and agricultural writer, John Seymour, "the British Isles have more breeds of sheep in them than the rest of the world put together." As their names (Southdown, Cheviot, Oxford, Suffolk, etc.) tell us, these different breeds were developed by local breeders in response to local economies, needs, and conditions. The existence of so many breeds in such a small area does not suggest that Britain somehow had the luck to produce a few agricultural geniuses, but rather that farmers and shepherds of independent, sound, sensitive intelligence—minds of a quality that would now be considered exceptional and rare—were fairly commonplace all over those islands for a long time.

It is impossible, of course, to give a full description of the mind of a good farmer. It is only possible to suggest—as I have tried to do—that the mind of a good farmer is comparable in complexity and quality to a good mind of any other sort. I have said enough, I think, to suggest that the work of a good farmer is not mindless, not simple, and that it therefore cannot be done by a simpleton.

The good farmer, like an artist, performs within a pattern; he must do one thing while remembering many others. He must be thoughtful of relationships and connections, always aware of the reciprocity of dependence and influence between part and whole. His work may be physical, but its integrity is made by thought. We will not understand what we mean when we say that he works with his hands, if we do not understand that he works also with his mind. The good farmer's mind, like any other good mind, is one that can think, but it is by that very token a mind that cannot in any simple way be thought *for*.

The good farmer's mind, as I understand it, is in a certain critical sense beyond the reach of textbooks and expert advice. Textbooks and expert advice, that is, can be useful to this mind, but only by means of a translation—difficult but possible, which only this mind can make—from the abstract to the particular. This translation cannot be made by the expert without a condescension and oversimplification that demean and finally destroy both the two minds and the two kinds of work that are involved. To the textbook writer or researcher, the farm—the place where knowledge is applied—is necessarily provisional or theoretical; what he proposes must be found to be *generally* true. For the good farmer, on the other hand, the place where knowledge is applied is minutely particular, not *a* farm but *this* farm, *my* farm, the only place exactly like itself in all the world. To use it without intimate, minutely particular knowledge of it, as if it were *a* farm or *any* farm, is, as good farmers tend to know instinctively, to violate it, to do it damage, finally to destroy it.

And so one of the reasons it is impossible to give a full description of a good farmer's mind is that the mind of a good farmer is inseparable from his farm; or, to state it the opposite way, a farm as a human artifact is inseparable from the mind that makes and uses it. The two are one. To damage this union is to damage human culture at its root.

To say that the good farmer's mind and his farm are an indivisible unit is to imply, among much else, that our present ruling definition of *property* is grossly inadequate, and it is to begin to say what is wrong with what most journalists and experts have to say about farming. If we define a farm as property, and property as a commodity that one owns and may therefore sell (that is, if we equate it with its monetary value), then we have entirely divided the farmer's mind from it. We have placed "the economy" between the mind and its work, because we have placed it between the mind and its workplace. And so it becomes possible to consider farming as a kind of economics and to welcome the failure of farmers who are not sophisticated businessmen.

This presents no problem for people who grant an absolute status to "the economy," seeing it as an infallible index of reality. In the August 1982 *American Spectator*, for example, Julian L. Simon wrote:

> Wonderful though this Illinois land is for growing corn and soybeans, it has greater value to the economy as a shopping center, which is why the mall investors could pay the farmer enough to make it worthwhile for him to sell. . . . Under these conditions, no one would ever argue that the land should be required to remain in the production of corn and soybeans.

If one believes that this is the best of all possible economies, then one obviously need not argue about *anything*. Things are as they are because the best of all possible economies has determined that they should so be. The questions of what might be the definitions of a good farm or a good mind or a good shopping center are not asked, and need not be asked, because in the best of all possible economies such questions are already answered: the good is what we have and it is what we are going to have. Similarly, if one grants this kind of standing to "the economy," it is impossible to ask if there can be an error of any kind other than economic. If the price is right, we know all we need to know.

Professor Simon's reasoning is attractively simple and relaxing. One would believe in it if one could. The difficulty is that it answers none of the questions that one would like to ask. It answers no questions at all. If one's thinking leads only to the conclusion that what happens happens and is good, one might as well not be thinking—which here seems to be the case.

If we are going to think about the economy, we are obviously going to have to stand something beside it to measure it by. And it is clear that money is too abstract—and far too elastic—ever to make a reliable measuring rod. If we should, on the other hand, make health our standard, then we will have to ask some questions about an economy that, by the logical extension of its assumptions, causes catastrophic soil erosion and eliminates millions of farmers from farming by the device of economic ruin. If we cannot look on these results and find them good, then we must feel that Professor Simon's economy is itself a problem that can only be solved by a better economy, able to take more into account. And we must suppose that a better economy would have to rest on a better understanding of property.

"Property," Chesterton said, "is a point of honor. The true contrary of the word 'property' is the word 'prostitution.'" By "prostitution" I assume he meant the neglect or abuse of what is properly one's own. And as Chesterton well understood, there are two kinds of prostitution: the personal kind, by which we disown responsibilities that properly belong to us, and the political kind—"a system equally impersonal and inhuman," Chesterton said, "whether we call it Communism or Capitalism"—which permits some people to define and tamper with responsibilities that belong properly to other people. Property, so conceived, is a mean between two opposite extremes, each of which tends to destroy both the meaning of the word property and the worldly goods to which it refers.

Property belongs to a family of words that, if we can free them from the denigrations that shallow politics and social fashion have imposed on them, are the words, the ideas, that govern our connections with the world and with one another: property, proper, appropriate, propriety.

The word property then, if we use it in its full sense (that is, with a proper respect for the pattern of meanings that surrounds it) always implies the intimate involvement of a proprietary mind—not the mind of ownership, as that term is necessarily defined by the industrial economy, but a mind possessed of the knowledge, affection, and skill appropriate to the keeping and the use of its property. This inseparability and reciprocity, this mutual dependence, of mind and place, implies invariably a propriety of scale that would limit the size of properties. To ignore or destroy this profound connection is finally to destroy property (and with it, of course, the human economy), for it is to destroy the kind of mind by which property is properly used and so preserved.

3 | Good Farming and the Public Good

Donald Worster

Rain is a blessing when it falls gently on parched fields, turning the earth green, causing the birds to sing. But when it rains and rains, for forty days and nights, as it did for Noah, then the waters rise and destroy. Life is everywhere like that. Too little is a curse, too much is a plague.

For thousands of years, the philosopher's task has been to discover an optimum point where men and women can live modestly and securely, avoiding the extremes. The philosopher may seek a point of environmental balance where there is neither too little nor too much of nature's gifts. Or he may try to define the point where private ambitions and collective needs are in harmony, where individual appetites do not overrun the common-wealth and society's demands do not cut too deeply into individual free-doms. When philosophy is applied to the definition of a society's welfare, we call that point the "public good." Farmers, more than most people, ought to be responsive to that philosophical quest for a harmonious, bal-

The author wishes to thank the Hawaii Committee for the Humanities for its generous sup-port in the writing of this chapter.

anced good, for it has been their aim over a long history to seek moderation from nature and cooperation from their neighbors.

Yet it has been awhile since American agriculture, as a whole, has enjoyed a feeling of balance. The problem has not been in nature so much as in our society. We have not had a feeling of balance because we have come to hold extravagant ideas of what agriculture should contribute economically to the nation and the farmer. These days we are not a people noted for moderate thinking, so perhaps we have no reason to expect the idea of moderate farming to thrive. The most serious consequence of an immoderate culture, I will argue, is that the public good will not be well understood and therefore will not be achieved—in agriculture or in other areas. Another consequence is that farmers in the aggregate will suffer immensely and so will the practice of farming.

That has indeed happened in America, and we can blame it on our extreme dedication to the goal of maximizing agricultural productivity and wealth. In turn, that goal stems from a larger cultural conviction that wealth is our main aim in this nation and that wealth is an unlimited good. Almost everything we have celebrated as our success in farming has been defined in terms of those ends. It has now, however, become clear that our ends have been our undoing. We wanted more rain, unlimited rain, and we got a flood. It has left American farmers drowning in dreary statistics: crop reports, production charts, mortgage rates, energy bills, land and commodity prices.

For the past century or so, the nation has had to deal repeatedly with gluts of farm production, particularly in the cash grains. In 1981, a U.S. Department of Agriculture study declared an end to that nemesis; farmers, it predicted, would see no more price-depressing surpluses because we were at last able to peddle all the excess overseas.[1] World population and demand had caught up at last with the American farmer. The study admitted that selling all our glut abroad would make the life of domestic producers more unstable than ever, that they would find themselves on a wild roller coaster ride, plunging up and down the track of international markets. But the old problem of recurrent overproduction had now been licked. There would never again be mountains of grain piled up and waiting for boxcars, never again be oranges or cotton or sugar cane going to rot.

Things have not worked out quite the way they were supposed to. In the summer of 1982, the old surplus problem appeared again: an all-time record wheat harvest of 2.1 billion bushels; not enough American buyers to absorb it; the Russians emerging once again as the main hope of salvation, but acting unpredictably as usual; crop prices declining; real farm income the lowest it had been in fifty years. Angry Colorado farmers, fighting against foreclosures on their farms, were tear-gassed by the sheriff, a replay of the turmoil in the 1920s and 1930s. Consequently, doubt begins to creep into our farm policy assumptions. Is there possibly something wrong with our approach to agriculture? How else can we explain the fact that farming, after so much attention, continues to be in so much trouble?

There can never be a perfect equilibrium in farming or any other sphere; existence might be unbearably dull if there were. But when you are flooded again and again, you are not in any danger of boredom. You naturally look up and ask, What's making so much rain? Can we turn it down a little? In America we have seldom accepted excess in nature; on the contrary, we have put our best talent and energy to work getting rid of it, making what is dry wet and wet dry. Why then do we accept excess when it is our own doing? Simply because powerful elements in our society do not allow us to recognize that there is such a thing as too much productivity, too much chasing after wealth. Despite overwhelming evidence that the idea is not working, good farming continues to mean *more*. We have not yet come to see that more of any good may, after a point, wash us away.

In the depressed 1930s, when times were even harder for farmers than they are today, American political leaders took counsel. They listened to ranchers, growers, sharecroppers, agronomists, soil experts, and marketing specialists; a few of those leaders raised basic questions of value. What, they asked, is agriculture for? What is the ultimate moral reason behind the pursuit of abundance, new farm technology, and an expanding economy—or is there one? Are our farm policies improved means to unimproved or unexamined ends? Whose ends should farming serve, those of the rich and powerful or those of the poor? How much of nature should we spend on human desires? What are rational limits to our demands? What is the public good in agriculture and what kind of farming will most likely achieve it? Those questions are as relevant today as they were fifty years ago.

To begin a reappraisal, let us consider what the public good has come to mean in the standard discussions of agricultural policy. "Cheap and plentiful food" is the most common theme. The public deserves to eat at the lowest possible cost, so we are told, and then they will be able to spend the rest of their wages on an automobile or at the movies or on college. This, in fact, is what our policy persistently has assumed to be the chief public interest. Agriculture is supposed to contribute mainly to the wealth of Americans generally, to make possible an ever-higher level of personal consumption.

Such has long been a typical farm expert's definition of the public good. But is it what the American public really wants? According to scattered public opinion polls, people say they prefer having lots of small farms around rather than a few big ones, more little dairies and truck gardens, not so many giant agribusinesses.[2] It might be that encouraging smaller farms would be worth more to people than saving a nickel on cheese or lettuce; a choice between the two options, however, is not clearly presented to them when they walk into a supermarket. The farm experts merely assume, on the basis of marketplace behavior, that the public wants cheapness above all else. Cheapness, of course, is supposed to require abundance, and abundance is supposed to come from greater economies of scale, more concentrated economic organization, and more industrialized methods. The entire basis for that assumption collapses if the marketplace is a poor or imperfect reflector of what people want. And it is. In matters of national defense, education, health care, and old-age assistance, we do not assume that the marketplace would be an adequate basis for public policy. Why then should we make that assumption in agriculture?

A corollary, and sometimes a rival, to the notion that good farming is farming that makes America richer through mass production, is the belief that farming is successful when it makes farmers as a special group more affluent. A common belief among policy makers is that swelling prosperity down on the farm immeasurably benefits society. No other sector of the economy has managed so fully as this one to identify its private fortunes with the public good—not dentists, not teachers, not even defense contractors. Billions of tax dollars have gone into scientific research and innovation to promote higher production, lower costs, and greater income on the farm. Heavily subsidized irrigation water flows to some very wealthy growers in the western states. Price-support programs, which have raised consumer prices by as much as 20 percent, have put money into agricultural pockets.[3] A huge department in Washington, as well as agencies in every state, looks out for the welfare of this group. Hardly anyone begrudges the farm sector all of these gifts, though it is well established that most of them go to the wealthy few. We have been taught from Thomas Jefferson's day on down that what is good for the farmer is good for America.

That doctrine has survived some revolutionary changes in the conditions of rural folk. Currently only about six million people live on farms, and only a small fraction of them (one in ten) now falls below the poverty line.[4] Contrast those numbers with the eleven million who were unemployed at the depth of the recent recession, roaming city streets, standing in bread lines, dreading the day their unemployment benefits ran out, or consider the several million more who have given up looking for work. Although farmers, like the unemployed and destitute, have at times experienced the indifference of other Americans to their plight, by and large they have retained an unusually sympathetic audience. When the man or woman on a tractor is in economic trouble, there is widespread worry; something, the newspapers agree, must be done. Farmers, particularly the better-off ones, maintain powerful friends and well-financed lobbyists to plead their case for governmental aid—unlike the city welfare mother who is roundly berated by the president and legislators in Congress, told to economize further on her food budget, to find a cheaper slum to live in, to forget a "handout" from Washington, to get off the public dole. In other words, the welfare and profit of this rather small group of largely middle- and upper-income farmers and ranchers have become identified with the public good in agriculture.

What is wrong with that identification? Do modern, business-oriented farmers not deserve our compassion when they go through hard times and are threatened with the loss of their assets and even their entire farms? They do, and they deserve our respect for their intelligence, hard work, fortitude, and skill. But respect and compassion should not be confused with favoritism or the private welfare of a single group with the larger welfare of the American commonwealth. When that confusion occurs, farmers are hurt; their welfare depends on a clear, reasoned concept of public good. What is good for all Americans, we must understand, will in the long run be good for farming, too.

The source of our difficulties is not a lack of popular concern for farmers or such superficial things as stagnant overseas markets or expensive credit. Rather, it is an inadequate idea of what truly constitutes the agricultural good of this nation.

The predominant idea of the public good in agriculture takes two forms: first, forever increasing the gross farm product and, second, forever seeking to augment the wealth of the farm sector (even if it means losing most of our farmers). Both programs are tied to the ideology and pressures of the marketplace. When pushed to the extreme as we have pushed it, that market mentality becomes seriously destabilizing to rural communities. It produces a perpetually crisis-ridden farm economy. Worse, it embitters people because it cannot deliver what it says it will: a general contentment and happiness. When the marketplace is made the main idea, it diminishes other values, leads to a degrading of personal independence, social bonds, virtue, and patriotism—for those qualities cannot thrive in an unbridled culture of acquisition, which the mentality of market maximization leads to.

In an earlier America of extensive rural poverty and poor living conditions more could be said for the vigorous pursuit of wealth in the marketplace, just as more may be said for it today in Bangladesh or Haiti. But when that pursuit persists beyond the point of material sufficiency, when it becomes a dream of unlimited economic gain, troubles follow. That is what has happened to American farmers and indeed to this country in general. Farmers must run their machines nonstop to keep up with the self-aggrandizing industrialists. The faster farmers go, the more crops they harvest, the more secure their position in the marketplace may be, the more they can buy—so they hope. But what they win in that way lasts only for a brief while. A continual uncertainty is their fate in this society.

The average farmer is not altogether responsible for this predicament. He did not set up the race, and he is not leading in it but is somewhere back in the pack, straining to catch up with corporate presidents, athletes, lawyers, movie stars, and engineers. The modern farmer lives in an intensely high-pressure world of many wealth maximizers. In that milieu, growing food becomes his only defense, his sole means of competing for social position. Unfortunately for him, food has been a comparatively poor basis for income growth, for it quickly saturates its market: humans can eat only so much lettuce or beef. Unlike others in the race, the farmer must always confront the biological limits of the consumer. He cannot make more money without finding more mouths and bellies to feed. Agriculture, by its very nature, is a productive activity that deals primarily with real human needs, not the contrived wants around which the game of maximization revolves. That difference must inescapably put the farmer at a disadvantage.

In another respect agriculture is not unique. It cannot evade the bitter disappointment over shrinking promises that is endemic in marketplace societies. All individuals cannot maximize their wealth; some people have to give up something in order for others to get all they want. The social philosophy of private accumulation is a lot like Calvinism: an elect few are chosen to live in paradise, while the rest can go to hell. The number of the elect is not fixed once and for all; it decreases steadily to the vanishing point. Especially since World War II that outcome has been a familiar experience in American farming. In 1900, there were 5.7 million farms in the United States, averaging 138 acres apiece. By 1978, the number had dropped to 2.5 million, and their average size was 415 acres. Over the past thirty years the typical American farm increased its spread by 20 to 30 percent each decade, and all of the increase was taken from a neighbor's side of the fence.[5] At that rate the promise of unlimited farm riches will someday soon be made only to a tiny privileged remnant.

Any farmer must look in the mirror each morning, like an anxious Puritan on his way out to church, and ask, "Am I to be among the saved or not?" Do what he will, the odds are clearly running against him. And if the promise of the land will tingle the ears of fewer and fewer people, if eventually it will belong to an oligopoly—and it is naive to believe agriculture alone can forever escape the corporate takeover—then the offer of unbounded wealth for farmers will turn out to have been a fraud.

The public good cannot be realized in agriculture, therefore, by the untrammeled workings of the market economy and the endless striving for private profit that it institutionalizes. The market creates wealth all right, but its wealth cannot satisfy; it holds up an ideal that is never really achieved, receding indefinitely before our eyes. A farm policy defined only in market terms inevitably must destroy the agricultural community to make it prosper. It must lead to disillusionment and frustration, uprooting and alienation, wearing farmers out, then casting them off.

That is not to say there are no benefits at all in the free-market approach; rather, the arguments on its behalf can take us only a short way toward locating the optimum good of our social life, and then they become immoderate and illusionary. It is now time in the United States to try another tack, to search beyond the marketplace to serve the public good.

Suppose we begin by simply asking what it is that we as a society want out of farming in the future. Are there significant human values that agriculture can help us to realize? Once we have answered that large question, we can call in the farmers, along with the agronomists, the economists, and the fertilizer salesmen, to make the ideal real. By that strategy we might establish better control over where we are going, decide where we want to end up, and stop the drift toward rural chaos. Good farming, by that approach, would be understood as the art, the science, and the wisdom of growing values in the soil—and no calling can be more honorable than that.

Slowly, several worthy answers to that large question of the common good in agriculture have begun to emerge in public discussions. They are familiar in one form or another to us all. The task now is to make them as compelling as possible and move them out from under the deadening shadow of profit maximization.

1. *Good farming is farming that makes people healthier.* It does so by creating and delivering food of the highest attainable nutritional quality and safety. Agriculture fails in its most obvious mission when that quality of healthfulness is missing or when it becomes corrupted by such things as toxic chemical residues. One of the most serious calamities to befall modern industrial farming is that it has turned food into a suspect, potentially dangerous commodity. When people begin to bite gingerly into apples, wondering whether cancer might be lurking there, or when they hesitate to

drink a cup of milk, remembering that heptachlor has been found in the dairy's cows, or when they are unsure whether chemical growth-stimulants are lingering in a chicken-salad sandwich, then agriculture has created for itself the most serious possible problem. After all, the essential point of farming is to keep people alive. No gain in export earnings or farm profit, no ease of harvesting or freedom from pests can justify risking human life, can excuse putting the public's health in danger; to act or think otherwise ought to violate ethics as much as the willful practice of bad medicine. Yet the willingness to risk life and health has become daily news in contemporary food processing and agriculture. The problem is compounded by the fact that farmers may conscientiously harvest crops that meet the strictest standards of nutrition and safety but then must turn them over to numerous processors who, for the sake of profit, have been known to take most of the nutrition out, put additives in, turn wheat into Twinkies and corn into breakfast-table candy.[6] The more complex and powerful the system of farm production, the more sensitive and strict must be the moral consciousness behind it and the more elaborate and expensive the system of public control overseeing it. There is no cheaper, simpler, easier way to realize this value.

2. *Good farming is farming that promotes a more just society.* For a long time in America, the land was where most people expected to go for their start in life, where they hoped to find opportunity and secure a living. The land, always the land: if not in this place, then farther west. Our society's thinking about fairness and democracy reflects even now a reliance on the land as an available, inexhaustible resource. Today, however, we are telling the majority of rural people that there is not enough farmland for them, that they will have to go someplace else for their livelihood, although it is never precisely indicated where that "someplace else" is. If agriculture passes the buck, where will it stop? Does agriculture not have an obligation to the poor and landless in its midst? An obligation to pay decent wages to its laborers and to make room for new farmers, rather than expecting the besieged, depressed cities to take the unwanted? Agriculture, through both private and public agencies, can and should give assistance to struggling racial minorities across the country: to black farmers who are living as tenants on worn-out land, to Indian farmers who need irrigation water, to small Hispanic growers who seek a fair share of attention from county extension agents, to Hawaiians who want land for taro and cultural survival. The agricultural community should work to lop the top off of the rural pyramid of wealth, which is reaching stratospheric heights; today a mere 5 percent of the nation's landowners control almost half the farm acreage, while in the Mountain West a minuscule 1 percent owns 38

percent of all agricultural land and, in the Pacific states, that same percent owns 43 percent of the land.[7] Agriculture, however, is not doing any of these things. On the contrary, it is everywhere retreating rapidly from a commitment to justice and democracy. Meanwhile, several other nations are managing, despite the pressure of the world marketplace and industrialization, to hold onto the democratic principle; the Danes, for example, have long pursued the ideal of a rural world where few have too little and even fewer have too much.[8] When our own farm experts and leaders rediscover that moral value, American agriculture will be stronger and more successful than it is today.

3. *Good farming is farming that preserves the earth and its network of life.* Obviously, agriculture involves the rearranging of nature to bring it more into line with human desires, but it does not require exploiting, mining, or destroying the natural world. The need for agriculture also does not absolve us from the moral duty and the common-sense advice to farm in an ecologically rational way. Good farming protects the land, even when it uses it. It does not knock down shelterbelts to squeeze a few more dollars from a field. It does not poison the animal creation wholesale to get rid of coyotes and bobcats. It does not drain entire rivers dry, causing irreversible damage to estuaries and aquatic ecosystems, in an uncontrolled urge to irrigate the desert. It does not tolerate the yearly loss of 200 tons of topsoil per acre from farms in the Palouse hills of eastern Washington.[9] Those are the ways of violence. American agriculture of late, pushed by market forces and armed with unprecedented technology, has increasingly become a violent enterprise.

Good farming, in contrast, is a profession of peace and cooperation with the earth. It is work that calls for wise, sensitive people who are not ashamed to love their land, who will treat it with understanding and care, and who will perceive its future as their own. Many farmers and ranchers are still like that and can give us all advanced lessons in ecological ethics. But most preservation-minded farmers are now old men and women, preparing for retirement. There is great danger that they will sell out to less informed, less careful individuals or corporations, who may acquire more earth than they can know intimately and farm well. Somehow we must avoid that outcome. Agriculture's future must be oriented toward using land according to the principles of practical ecology—toward a conserving ethic and intelligence. That orientation is essential if we want to leave our children a planet as fruitful as the one we inherited.[10]

Other public goods I have not mentioned include creating beauty in the landscape, strengthening rural families, aiding the world's hungry—especially helping them produce their own food, diverting investment capital into other sectors of the economy that now need it more than agriculture, and preserving the rural past and its traditions. All of them require us to make policy, not merely make more food. These common goods do not assume that the lot of farmers can be bettered without also considering what the entire human community requires.

Americans are often accused of being a privatizing people. We take the question of public good and break it into millions of little pieces, into every individual's private wants, and then reduce it further by trying to put a price on those pieces. This is my property, we say, and no one can tell me how to farm it. These are my cows, and I will graze them where I like and as hard as I like, until the grass is dead if I like.

There is another America, however, one that has been more open to ideas of the common welfare. That America usually can be found today in less progressive corners, often in rural neighborhoods where there is still a long memory running back to a time when farm folk got their living together and worked more as partners with the land. The future of agriculture will be determined by whether that community thinking can be nurtured and grown more abundantly. It is easily the most important crop we can raise and harvest in the United States.

Those who have forgotten what that sense of rural commonality was like in earlier periods can find it again in the pages of history and literature: in a novel, for example, like O.E. Rölvaag's *Giants in the Earth*, which depicts the settling of the Dakota prairie by a group of Norwegian immigrants.[11] They brought little into that grassland besides themselves and their old-country habits of mutual aid. When nature gave them more than they wanted—gave them plagues of grasshoppers, droughts, and blizzards—they struggled together and came through to more moderate times. They endured and even prospered, but they did not become rich. They made homes for themselves, but they did not conquer that hard land. What they achieved was a wary peace with the prairie, an affectionate and understanding peace, a peace that reflected the fact that they were at peace among themselves. Each family had its own property and fences, its own way of doing things. Occasionally they competed against one another in friendly rivalries. But the overarching principle of their lives, as Rölvaag describes it, was the maintenance of a social bond, which finally became a bond with the strange, foreign land where they settled. Communalism of that sort, in real life as well as in fiction, is receding today, but it has not yet altogether disappeared over the horizon.

The challenge now is to retrieve that commitment to community from the past, from scattered pockets of rural life, and to find a modern expression for it in this new age of industrial agriculture.

At the heart of any nation's agricultural policy must be its ideal of a good farmer. For a number of years we have told farmers, through our colleges, agricultural magazines, government officials, and exporters, one clear thing: get as much as you possibly can out of the land. We have not told them how many farmers would have to be sacrificed to meet that instruction or how much it would deprive the few who remained of their freedom, contentment, or husbandry.

But sooner or later the prevailing ideals wear out, giving way to new ones or to new versions of even older ones. The American ideal of good farming, and the agricultural policy we have built on it, may be ready for a shift in the directions suggested here. In the not-too-distant future, farming may come to mean again a life aimed at permanence, an occupation devoted to value as well as technique, a work of moderation and balance. That is a shift in which we all have a stake.

4 The Making and Unmaking of a Fertile Soil

Hans Jenny

When the early immigrants settled in Pennsylvania, they were astounded by the high fertility of the valley and limestone soils.[1] "Crops are bountiful and livestock is thriving," wrote the Mennonites to their brethren in the old country, where good land was scarce and, in some places, a vacating tenant farmer could take topsoil to a new location. In the New World, instead, manuring and liming did not pay. Soon the rules of European soil husbandry were forgotten, except in tradition-motivated communities such as those of the Amish. Many pioneers acquired more land than they could handle, and soil washing—that is, erosion—became rampant. People shrugged it off: when soil gives out, more is available out West.

When the stream of immigrants reached Ohio and the Middle West and cultivated the rich, silt loam soils, the local surplus of food became staggering. To move it to markets, fattened cattle and hogs were driven in herds

The author acknowledges valuable comments from his colleagues R. J. Arkley and A. M. Schultz; former graduate students D. Maher and J. Sandor; L. R. Wohletz, former Soil Conservation Service (SCS) state soil scientist; and advice and information from SCS staff at California headquarters. All judgments and suggestions are those of the author.

across the mountains to Boston and Philadelphia, and corn and wheat were loaded on to Mississippi barges destined for New Orleans. Transferred to sailing ships, the cargoes headed for Europe. To cope with this flood of cheap grain, a small country like Switzerland had to shift from wheat farming to dairying and cheese making.

No wonder America's fertile soils became legendary abroad and attracted endless caravans of newcomers. And to this day, American agriculture produces an abundance of food that is declared at the highest political levels as an effective international weapon. But what is happening to the soil resource? The answers weave a tale that should concern a wide American audience.

MANY KINDS OF SOILS

Since pioneer days, many thousands of productive and unproductive soils, colored black, gray, brown, yellow, white, olive, pink, and red, have been mapped, described, classified, and interpreted for use. Each soil is an individual body of nature, possessing its own character, life history, and powers to support plants and animals. The classic examples of superior soils are those of the native midwestern prairies that nourished bluestem grasses tall enough to hide a rider on his horse. At the other extreme, visitors are appalled at the expanses of barren, deep serpentine soils on the slopes of San Benito Mountain in California. Not a blade of grass is seen, not a weed, not a flower. Between these extremes are the multitudes of ordinary soils supporting good forests and grasslands, and a prosperous agriculture when the times are good, but having impediments of one sort or another, such as being shallow, erodible, too steep or too flat; having too much sand, clay, salt, or acidity; or suffering from subsoil cementation (hardening of the soil), poor internal drainage, or a high water table.

Study of the fate of American soils is facilitated by understanding how soils are formed in the first place. We shall trace nature's creation of a fertile soil in the temperate region and observe its deterioration in our society.

A LIFELESS BEGINNING

When a volcano erupts, as Mt. St. Helens does nowadays and Mt. Vesuvius did in Roman times, thick covers of ashes and hot lava flows play havoc with the countryside. Within five years, or ten, greenery appears in spots and life resumes. When alpine and Alaskan glaciers melt and retreat, large heaps of boulders, gravel, sand, and rock flour are left behind, posing a dreary scene in grays; yet in a few years or decades, patches of green develop. Either group of deposits is the parent material of a soil and either situation is a starting point or initial state of genesis. The processes that mold a soil may last for centuries, millennia, even millions of years, and the various evolutionary stages are occupied by transient communities of plants and animals.

Locked up in the sterile rock materials are most of the two dozen nutrient elements, such as phosphorus, sulfur, calcium, and potassium, that are needed by organisms for growth and survival. The fertility of a soil is strongly conditioned by the chemical makeup it acquired at birth; volcanic ash deposits, for example, are superior to coastal sand dunes that are high in quartz grains.

THE NITROGEN CRUX

Although winds carry insects, small seeds, and microorganisms to the new landscape, little growth ensues because the parent materials lack the element nitrogen, which is a key ingredient of the proteins, the giant molecules that are the staff of life. Small molecules of nitrogen gas are abundant in the air we breathe, but neither we, nor the animal world, nor the green plants (save the bluegreen algae) are able to utilize them in body metabolism. Just a few kinds of soil bacteria and other soil microbes possess the biochemical machinery for fixing air nitrogen and using it to construct their own body proteins. To secure the fuel needed for breaking the nitrogen molecule, the bacteria consume and digest carbohydrates such as sugars and starches manufactured by green plants. When some of the master bacteria die, other soil microbes dismantle the protein molecules and convert part of the nitrogen to ammonia and nitrate, which are preferred plant nutrients.

The linkage of nitrate production and sugar production is an obstacle to building a fertile soil, for plants cannot grow unless the substrate provides nitrates generated by microbes, yet microbes cannot deliver nitrates unless green plants furnish sugar and other carbohydrates. And neither of these basic operations can proceed without the soil supplying magnesium, iron, phosphorus, molybdenum, and other elements to the enzymatic reaction sites. In practice, the biotic inertia is slowly overcome by leaves and germules (seeds and spores) blown in, small amounts of ammonia and nitrates arriving in rainfall, and by the remarkable bluegreen algae fixing both carbon and nitrogen.

Nature devised a shortcut for privileged situations. Some of the nitrogen-fixing bacteria learned to live in small nodules attached to roots of lupines, clovers, and other leguminous plants. The bacteria receive sugars from the host and deliver nitrogen to it in return. A few brush species and alder trees also cultivate microbial friendships or symbioses. Patches of nitrogen-fixing plants quicken soil formation by enhancing input of organic matter and mobilizing rock-calcium which stabilizes cell walls and soil pores.

Nitrogen as a limiting element comes to the fore again when humans take charge of soils. In modern agriculture, nitrogen is commonly the first element needing replenishment by fertilizers or legumes.

BREATHING LIFE INTO THE SOIL

Once the soil is clad in a green carpet, organic matter is being added in self-propelling fashion. More roots penetrate the soil, and they reach greater depths. Litter from grasses, shrubs, and trees offers a wide spectrum of organic molecules, a smorgasbord for tiny organisms. Looked at under the microscope, the soil is teeming with millions of bacteria and molds. Hordes of each of a thousand species of small animals seemingly arrive from nowhere. In the eternal night of the soil, eyes and coloring offer no advantage, hence many of these animals are blind and colorless, but endowed with refined senses of hearing and smelling. The ants, spiders, springtails, sowbugs, centipedes, and worms endlessly push particles up and down, dig tunnels and channels, and arrange the soil framework to a livable abode.

In sunlight, green plants put together water from the soil and carbon dioxide gas of the air, making carbohydrates and oxygen gas. Plants are the carbon fixers, the producers of food for microbes, animals, and humans, who in turn are the decomposers, consuming oxygen and carbohydrates and breathing carbon dioxide back into the air. Not all that is fixed is being decomposed. Carbon is stored in tree trunks and massive roots as wood and in surface soils and subsoils as large, black humus molecules. These are cardinal soil ingredients because of their nitrogen content and chemical reactivity. By trapping toxic metals, such as lead and radioactive plutonium, and many herbicides and pesticides, humus acts as a sanitarian.

The contract between soil and vegetative cover is a kind of superorganism, a living system, called an ecosystem. Bare soil itself is an ecosystem of minerals, organic substances, and life forms. Vigorously, it inhales oxygen gas and exhales carbon dioxide gas, and, by analogy, nature writers think of such a soil as being in good health.

PUTTING MORE HUMUS UNDERGROUND

As long as plant production outpaces consumption by the decomposers, soil organic matter—with its carbon, nitrogen, and humus contents—keeps accruing. Throughout this long period, soil functions as a sink for

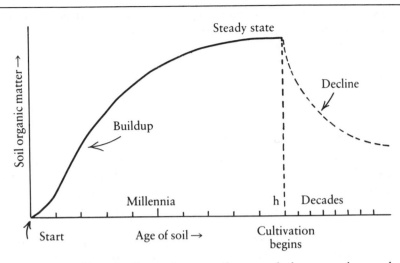

FIGURE 1. Buildup of soil organic matter (for example, humus, carbon, and nitrogen) in nature during millennia and destruction during decades of farming. At "h," cultivation begins.

carbon. Eventually, annual buildup and decay balance each other, and soil and natural vegetation become stabilized and remain in a more or less balanced or steady state for a long period of time. To a depth of one meter, a rich, virgin soil may contain 450 tons per hectare of organic matter, mostly humus, and 17 tons per hectare of nitrogen, not counting that in roots and surface litter. Based on observations of soils of known ages, the solid curve in Figure 1 portrays the trend of organic matter accumulation in soil for periods of up to 10,000 years. Nature's buildup is slow, about ten to twenty kilograms per hectare per year of nitrogen in the early stages of fertile soils.[2]

THE FERTILE SOIL

As soils develop, the initial material becomes strongly modified. Changes in topsoil and in subsoils—that is, in the soil profile—are revealed along road cuts and, better still, in a freshly dug pit that displays the often contrastingly colored and sculptured layers or horizons. The ebony black surface soil high in humus, as found near the Canadian border, fades to light gray and ivory in the lower part of the subsoil and mirrors the distribution pattern of grass roots. Iron is common in soils and oxidizes to rustiness, endowing many soils with reddish brown tints. In Hawaii's tropical climes, rocks rich in iron underlie a dark-red soil mantle.

Fortified with carbonic, nitric, and sulfuric acids generated by soil microbes, the soil solution enhances weathering of rock particles, thereby liberating nutrients and rearranging silicon and aluminum atoms to tiny hexagonal, plate-shaped clay particles. Clays are bonded with humus molecules and form stable crumbs that loosen the soil and facilitate aeration. When a balanced assembly of sand, silt, clay particles, humus globs, rootlets, and small organisms—the entire soil material—is crushed between moist fingers, it feels like a mellow loam. Because they have good tilth and favorable porosity that stores rainwater and drains excess water, loams are highly esteemed.

This soil, as envisioned here, is fertile because it is free of toxic substances. It is one to two meters deep and lacks root-restrictive subsoil conditions. Further, analysis of the soil solution and of the nutrients held by clay and humus particles indicates good supplies of phosphorus, potassium, calcium, and other essential elements, and incubation records high rates of nitrate delivery. In nutrient cycling, phosphorus, potassium, sulfur, etcetera, are transported from roots in the subsoil to stems and leaves and are returned in litterfall, thereby enriching the topsoil, which becomes the most valuable part of the soil profile as far as agricultural interests are concerned.

A fertile soil is also a productive soil whenever rainfall, temperatures, and other environmental influences are conducive to growth. A further definition is needed. As used here, *soil* means a soil profile of variable depth. A pile of sand and a bucket full of surface soil are mere soil materials, though loosely, they could be called soil.

SOIL OBLITERATION

Water-laid soils along the American River in California are fertile and support agriculture. They contain small quantities of gold dust, which is their undoing. Mining companies buy up farmland and extract the gold by sieving and chemical means, thereby turning the soil upside down, leaving heaps of large cobbles on top and washing out the clay.

On the corporate books, the operation is a success. New wealth is created, pounds of gold that contribute to the gross national product. The destruction of land that could produce food for countless generations does not appear on the ledger. And the hidden social costs are not considered; some of the bought-out farmers are not used to handling large amounts of cash, and before long their families are on the welfare rolls.

Land has to be sacrificed for buildings, towns, and cities. Modern society devours land with alacrity. More housing projects, boulevards, roads, and highways are built, more shopping centers with larger parking areas, more airfields, military installations, factories, and corporation yards are constructed. Soil is dug up, moved around and covered with thick layers of concrete; it ceases to be a biological producer. Urban sprawl prefers first-class farmland because it is level and well drained, hence cheaper to build on. For the entire area of the United States, the annual loss of farmland amounts to less than 1 percent, but in terms of the area of first-class land, the disappearance is alarming in many states. Recreational and wilderness areas also withdraw land from food and fiber production, but in an emergency, the good soils are readily converted to agricultural use.

Well-meaning people tell us that soils are no longer needed because scientists can grow plants in artificial nutrient solutions, and sunshine may be replaced by light bulbs. We are not told how many generations of a given plant variety can endure the artificial environment or what colossal and costly installations will be needed to feed millions, provide pastures, and raise fruit trees.

SOIL DESERTIFICATION

When Hun Attila's Asian warriors moved westward, he cowed his enemies by threatening to throw salt on their lands. Salt—salinity—and alkali infestations are still menacing the many populations that live in arid and semiarid plains and valleys.

In natural drylands, the water tables are often salty, but stay at great depth. The soils above are salt-free and support grasses and legumes whose stature is small because of the paucity of water. When irrigation canals are installed, the fresh water gratifyingly escalates the crop yields and the costly enterprise is declared a success.

Over the years, leaks in canals and excessive irrigation send water below the root zone and the groundwater table slowly rises. Long before it reaches the land surface, the wicklike suction of the soil will have accumulated enough salt to kill all crops. White salt incrustations, black blotches of alkaline humus, and scattered, salt-tolerant bushes and herbs take over. The dreaded "rise of the alkali" is one of the several manmade forms of desertification.

Where salt and soil chemistry are compatible and suitable topographies prevail, reclamation of sands and loams, but rarely of clay soils, has been accomplished in tedious, costly steps. The water table is lowered by year-long pumping or by tile drainage; chemical soil treatment is initiated, and salt mixtures are leached out over long periods of time. Certain lands may be restored to full productivity.

SOIL PUDDLING AND COMPACTION

Decades ago, arid land agriculturists entertained the belief that weekly plowing and disking of orchards conserved soil moisture. Lengthy research proved the water savings to be small and the soil damage to be great. Soil clods continually hit by metal spikes and blades lose the weak bonds of aggregates, and the soil becomes fluffy and dusty. Wetting and water movement are impaired. Fortunately, cessation of cultivation over a period of years, combined with cover cropping, gradually eliminates the puddled state of the surface soil.

Heavy steel plows leave a narrow zone of compaction, the plow sole, at the bottom of the furrow. The compressed layer reduces water penetration and curtails air supply to deep roots. The mere weight of heavy tractors, wagons, and sprayers compresses the subsoil, which reduces crop yields. Whereas one day of irrigation may be required to wet a good soil to the depth of 1.25–1.50 meters, compaction doubles and triples the time span, upsetting the operator's irrigation schedule and those of neighbors. In orchards that receive animal manure and grow cover crops to be turned under, compaction horizons (layers of compacted soil) caused by heavy traffic during disking, pruning, spraying, fertilizing, and harvesting are reduced in density but not eliminated.

SOIL MINING

Soon after American forests and grasslands were cut and put to the plow, astonishingly high crop yields were often reported. Farsighted experiment stations analyzed the virgin soils and kept track of the changes occurring under different systems of farming. For level parcels receiving no fertilizer and suffering no erosion, the soil's content of organic matter declined continually—rapidly during the first few decades, slowly afterwards. Crop yields also diminished. Under average farming conditions, over one-third (35 percent) of the nitrogen and carbon content had been eliminated in the first fifty years. In a prairie soil in Missouri, the actual loss in humus amounted to thirty-one tons per hectare. Labile, dynamic portions went first; resistant and inert ones stayed.

The trend is sketched in Figure 1 as a dashed, declining curve that may approach a new steady state, a much lower one than the virgin one. What nature had built up during thousands of years, farming dissipated in a few decades.[3] (In parts of India, over two-thirds of the humus content have been lost in centuries.)

Two causes account for the rapid oxidation, or burning out, of humus during conversion. First, nature's return of dead grasses and foliage ceases, except for leftover crop trash; in addition, plowing and stirring the soil enhance the decay activities of the decomposer organisms, leading to a reversal of the buildup curve. As would be expected, the nutrient holding capacity of the soil is lowered, mineral nutrients are lost, soil acidity rises, the crumbs break up, and soil structure and tilth deteriorate, making the soil less friable.

Types of cropping modify the losses; corn growing is the most abusive, pasturing and haymaking the least. This variation suggests that the decline curve might be arrested at any point by the choice of farming practices.

SOIL FERTILIZATION

Wood ashes, land plaster (gypsum), marls, and bone meal are old standbys of soil improvement. Maverick farmers would spread mixtures of crushed rocks over their lands. Peasants used to manure their impoverished fields with leaf litter collected from woodlands that became stagnant. They practiced the maxim of robbing Peter to feed Paul, which is still in vogue.

By the middle of the past century, the notion that wood ashes were useless plant impurities was displaced by the realization that they contain essential nutrient elements. Potassium (K) salts, long the spoils of table salt mining, suddenly appeared valuable. Fossilized bones, or rock phosphate, made soluble with sulfuric acid, turned into an available source of phosphorus (P). The invention of industrial nitrogen (N) fixation during World War I prolonged that conflict because it enabled our opposition to produce more gunpowder and also more food.

The big three, NPK, helped revolutionize agriculture. Soils that had been cropped for centuries could be rejuvenated, and mediocre soils, the stepchildren of nature, achieved respectable yields. Alas, chemical cures leave bad side effects: high dosages bring diminished returns and waste nutrients; high yield and high quality may not go together; plants high in nitrogen attract pests; potassium and ammonium salts acidify the soil and release toxic aluminum to the soil solution; when more nitrogen is given to a field than plants and microbes can utilize, the excess may pollute ground water and springs with nitrates, endangering the health of humans and animals. Wet spells convert nitrates to nitrites and to nitrous oxide gas, making things worse. Nitrogen fertilization is an intricate plant-soil operation that deserves to be monitored by trained personnel. High dosages of nitrogen in organic form lead to similar consequences.

always a balance necessary

In contrast to nitrates, phosphates react chemically with the soil and become less and less available. They are being locked up. The world's resources of rock phosphates are localized in North Africa, China, and the United States and they are expected to last for centuries, but not if Third World nations demand their share. Wars for phosphate deposits are not unthinkable, unless means are discovered to mobilize insoluble soil phosphates, as is claimed to be accomplished by the symbiotic fungus Mycorrhiza.

Is the current heavy use of NPK improving the soil's humus capital? It would depend to what extent fertilization enhances root growth, on one hand, and stimulates decomposer organisms, on the other. Without the major share of the produce being turned under, the balance sheet for soil organic matter may well be unchanged or negative, as is indicated by cultivated Missouri fields analyzed in 1930 and again in 1980. The spectacular advances in agricultural crop production have left these soils impoverished.[4]

CONFRONTING SOIL EROSION

As gentle rains continue falling on forested or grassy slopes, water soaks into the soil and fills it to capacity. Excess water is transmitted to lower strata. In heavy downpours, the raindrops arrive en masse and a portion of the water flows downhill as surface runoff. Though it is practically colorless, the water carries small amounts of clay and humus particles contributed by worm casts and gopher mounds. This type of *geologic* or *normal erosion* may lower the soil surface by less than a millimeter in a century, yet it carves new landscapes in millions of years.

Once land is cleared and leaf litter and grass mulch are gone, the raindrops hit the bare soil surface with full force, break up crumbs and granules, and initiate *accelerated erosion*. The fragments seal pores and the fast-flowing runoff waters are muddied with silt and clay. In upper portions of a slope, dark, fertile topsoil layers and light-colored subsoil as well are washed off and redeposited on the flatter foot slope below. Soil transport causes the ruinous *sheet*, *rill*, and *gully erosion*. Part of the suspended mud follows waterways to clog channels, lakes, and reservoirs and damage valley lands. A small part reaches the large rivers and, summed up over a continent, billions of tons of good soil material are delivered to the seas. Individual fields experience annual losses up to 200 metric tons per hectare; this high figure corresponds to dozens of heavy truckloads of soil material. For the southern Piedmont Plateau, overall soil truncation is estimated at 18 centimeters since white settlement began. Much cotton and tobacco are now grown on red, tough subsoil.[5]

The Soil Conservation Service, inaugurated during Roosevelt's New

Deal regime, has made heroic efforts to help farmers combat the ravages of erosion. The early enthusiasts believed that farmers would abandon their wasteful ways once shown that crop rotation, strip cropping, contour plowing, and grassy waterways pay off in the long run. Regretfully, only half of the farmers request conservation plans and only half of the applicants carry them out, in spite of decades of educational persuasion and billions of dollars in congressional appropriations.

How to appraise the efficiency of means of erosion control raises highly technical and sophisticated questions, baffling many good minds. Customarily, runoff waters from plots or small watersheds are analyzed; one might think that zero erosion is the desired goal. Since, however, natural ecosystems at steady state experience slight normal erosion—which is, assumedly, balanced by soil formation—the U. S. Department of Agriculture tolerates annual erosion rates up to 11.2 metric tons per hectare (5 tons per acre), believing that nature creates that much new soil each year. This magnitude of soil formation is of cardinal scientific interest, but it is not clear whether rock weathering, clay formation, or humus accumulation were monitored or how it was done. Sometimes deep soils are allowed higher erosion rates and shallow soils lower ones. The tolerance values invite apprehension because they might be too high and lead to excessive erosion. One millimeter of soil truncation a year does not seem much, but if we want our nation to remain the Promised Land for at least a thousand years—which is not long by Old World time scales—the tolerance values appear intolerable. Safe rates are those of normal, natural erosion, which are a hundredfold lower, as listed in Bennett's tabulations.[6] As a gratuitous benefit, the application of natural loss rates to soil management would assist in preserving the beauty of the American agricultural landscape. Fortunately, a recent publication by the Soil Science Society of America takes a fresh look at the erosion tolerance problem.[7]

Many workers are crop conservationists rather than soil conservationists because their primary concern is maintenance and enhancement of crop yields. They do not worry much about soil erosion as long as chemotherapy of soils with NPK keeps filling boxes, bins, and silos. They seem to underestimate the importance of subsoil depth for potential nutrient supply, water storage, root discharge, and root anchorage, which in the case of walnut orchards extends to twelve feet underground. Their conservation position includes the engaging and elusive goal of "preservation of the inherent productive capacity of the soil."

To explain, crop yields register the productivity of an entire agricultural ecosystem composed of soil, plant varieties, density of planting, weed and pest control, irrigation, fertilization, cultivation, length of growing period, mode of harvesting, and, last but not least, the prevailing weather. In field work the partial contribution of the soil to yield cannot be sorted out and, at best, only differences can be identified, as in temporal yield changes of a given soil whose other production factors remain the same.

In practice, the constellation of growth factors is covering up soil degradation by heightened yields. Soil decline has to be evaluated by field examination of the soil itself, including follow-up work in laboratories and greenhouses. Soils deserve having standing in their own right; their role is broader than producing food. To cite just two examples, soils maintain the water household of a landscape and contribute to aesthetic experiences.

SOIL RESTORATION

Conservation aims to save what is left of the soil resource—a desirable but limited goal. If better and higher yielding plants are to be grown in the future, the soils must be more supportive, too. Looking back, we remember that this country once had more fertile, stable, and resilient soils. The time has come to consider restoring the soil, rebuilding the soil body.

North America is unique among continents in having vivid memories and scientific documentations of what its agricultural lands looked like when white people displaced the Indians. Taking the near-virgin state as a reference point will indicate in which directions and at what rates our fertile soils are drifting. Rehabilitation need not imitate nature's steady state, which is seen as a benchmark in a changing soil landscape.

Soil washed into lakes and seas is irretrievable, but some mineral soil material might be moved uphill to fill vicious gullies and rilled fields. Tolerance values would be set near zero. Restoration of the humus capital and concomitant rehabilitation of crumbs and soil architecture may be a reversible process, provided lost mineral nutrients are replenished by judicious fertilization or imported in organic form from other areas. Organic crop residues should stay on the land and be returned to it from cities and towns as feedlot manure, sludge, and septage, although the uncertainty of metal toxicity needs clearing up. The much publicized conversion of organic farm and forest debris to fuel energy is not in the interest of quality maintenance.[8] Stimulation of root growth and attending humus formation might be achieved by breeding plants that transfer more carbohydrates from tops to roots than is current now.

A promising activity in erosion control and soil restoration combats the great soil skinners—corn, cotton, and soybeans grown on plowed, bare land. In the new, no-till system, the plow is discarded and stem and stalk residues of the previous crops are left on the ground as mulches that curtail water and wind erosion. The new crop is seeded in narrow furrows cut through the mulches. Absence of tillage favors weed growth, which is controlled by herbicides that pose financial and environmental risks. Wes Jackson's revolutionary proposal of abandoning annual crops altogether and growing instead perennial corn, wheat, cotton, etcetera, comes closest to nature's benevolent ways, but must await long-time work in plant breeding.

How far restoration should be pushed toward the soil's virgin state is problematical; whatever point is chosen on the dashed curve in Figure 1, the effort will be expensive. Many urbanites take it for granted that farmers are anxious to assume stewardship of their land. Some are. Others, burdened by debts and mounting doctor bills, cannot be expected to husband their soils for the benefit of future generations. If society takes the long view, which is in its interests, it will have to foot the bill for assurance of a continued food supply for home consumption and shipments abroad.

City dwellers facing overloaded grocery shelves and reading about food surpluses are not interested in agricultural woes, and to most farmers the scenario might seem to be utopian. A modern stance of inaction invokes the wonders of science and technology that will solve agricultural problems with well-timed breakthroughs. Government financing of soil protection is opposed by those economists who foresee drastic food shortages that will render farming lucrative, soil rehabilitation attractive, and urban sprawl prohibitive. Either plan envisages a higher portion of earning to be spent on acquisition of food and fiber than is current now. I, for one, prefer to pay for preservation of good land now rather than to pay later on for scarcity and infertility of land.

SOIL MYSTIQUE

The word *soil* is not particularly pleasing to see or hear, at least not in comparison with the European equivalents "*sol*," "*suelo*," and "*Boden*," words that lack any negative connotation. Our national, undignified attitude toward soil deserves a change, exemplified by the student who finished his term paper with this sentence, "Never again shall I call soil dirt."

Many of us derive pleasure from touching and handling soil or sitting on bare ground and sensing its earthy fragrance. Impressed by soil colors and textures, people have named their habitations after soils: Redlands, Red Bluff, Black Earth, White Sands, Adobe Meadows, etcetera. Famed Iowan Grant Wood painted in his prairie landscape *Arbor Day* a dark-brown soil profile with stylized horizons as a symbol of fertility. Frenchman Jean Dubuffet attaches to his soil paintings such labels as *The Humus, Voice of the Soil, Celebrations of the Soil, Secrets of the Subsoil,* and *Poem of Iron Oxide.*

Plowing the soil signals the onset of a new cycle of organic renewal. A fertile soil sustains people and purifies buried, dead bodies of their pathogens. In older cultures, soils are esteemed, even venerated. Soil is Mother Earth. Healthy soils make healthy people. During religious rituals and in states of emotion, people kneel and kiss the ground. Wars have drenched soils with blood, and people remember it. From the soil we come and to the soil we return. It all adds up to a soil mystique that transcends the notion of soil as a mere economic commodity or impersonal object of science.

THE FATE OF OUR SOIL

Many people believe that the human species was created last and that the earth, its animals and plants, are here to serve human purposes. To subdue nature is lauded as a noble deed. America's tradition of manipulating its natural resources for private gain, compounded by short-range policies, has been disastrous for many soils. Anticipation of the future fate of the soil resource is likewise discouraging because soil malpractice is rising. Although we cannot foresee the wants of American society a century from now, we know that—barring a nuclear holocaust—people will have to eat and that food will be grown on soils left over from our own soils. The coming century is the lifetime of our children, grandchildren, and great-grandchildren; it will be a time of fast population growth and of predicted global warming that is expected to push deserts north and the corn belt into Canada. And yet, as N.A. Berg, a former chief of the Soil Conservation Service, has pointed out, "There is no national policy for the United States that places value on the soil resource or on conserving the soil resource," and none is in sight.[9]

5   *Thinking Like a River*

Donald Worster

When we drive by a modern farm, we still expect to see green plants sprouting from the earth, bearing the promise of food or cooking oil or a cotton shirt. Pulling up one of those plants, we are still prepared to find dirt clinging to its roots. Even in this age of high-tech euphoria, agriculture remains essentially a matter of plants growing in the soil. But another element besides soil has always been a part of the farmer's life—water. Farming is not only growing crops on a piece of land, it is also growing crops in water. I don't mean a hydroponics lab. I mean that the farmer and his plants inescapably are participants in the natural cycle of water on this planet. Water is a more volatile, uncertain element than soil in the agricultural equation. Soil naturally stays there on the farm, unless poor management intervenes, whereas water is by nature forever on the move, falling from the clouds, soaking down to roots, running off in streams to the sea. We must farm rivers and the flow of water as well as fields and pastures if we are to continue to thrive. But it has never been easy to extract a living from something so mobile and elusive, so relentless and yet so vulnerable as water.

If there is to be a long-term, sustainable agriculture in the United States or elsewhere, farmers must think and act in accord with the flow of water over, under, through, and beyond their farms. Preserving the fertility of the soil resource is critical to sustaining it, of course, but not more so than maintaining the quality of water. In many ways, the two ideals are one. And their failure is one, as when rain erodes the topsoil and creeks and rivers suffer.

But there are differences between those two resources, differences we must understand and respect. Unlike soil, water cannot be "built." It can be lost to the farmer, or it can be diverted, polluted, misused, or over-appropriated, but it can never be deepened or enhanced as soil can be. There is only so much of it circulating in nature, and then there is no more. A sustainable agriculture is one that accepts and works carefully within the firm limits of the water cycle. It is one that exercises restraint in the demands it makes on the cycle. It is farming as close as possible to the nature of water, flowing with the current rather than opposing or obstructing it, as the Taoist philosophers of China recommended a long while back.[1]

Throughout history, the water cycle has served humans as a model of the natural world. Early civilizations saw in it a figure of the basic pattern of life, the cycle of birth, death, and return to the source of being. More recently, science has added to that ancient religious metaphor a new perception: the movement of water in an unending, undiminished loop can stand as a model for understanding the entire economy of nature. Looking for a way to make the principles of ecology clear and vivid, Aldo Leopold suggested that nature is a "round river," like a stream flowing into itself, going round and round in an unceasing circuit, going through all the soils, the flora, and the fauna of earth.[2] Another scientist, Robert Curry, has argued that the watershed (the area the river drains, its body, as it were) is the most appropriate unit for thinking about and dealing with nature.[3] The watershed is a complex whole, uniting biota, geochemistry, and energy in a single, interdependent system, in a dynamically balanced set of countervailing forces—erosion and construction, productivity and grace. Each watershed has its own peculiar shape and its own way of moving toward an elegant equilibrium of forces. The language of these scientists may be novel, but the insight is old and familiar. In water we see all of nature reflected. And in our use of that water, that nature, we see much of our past and future mirrored.

The first commandment for living successfully in nature—living for the long term at the highest possible level of moral development—is to understand how that round river and its watershed work together and to adapt our behavior accordingly. Taking a purely economic attitude toward water, on the other hand, is the surest way to fail in that understanding. In a strictly economic appraisal, water becomes merely the commodity H_2O, bulked here as capital to invest some day, spent freely when the market is high. It comes to be seen as a "cash flow," no longer as the lifeblood of the land. And then we begin to do foolish things with our streams and rivers. We fail to see that the meanderings of a creek across a meadow exemplify not chaotic or wasted motion but a fundamental rationality, that those meanderings make sense. Government engineers, confident that they know better, straighten the creek with bulldozers so that it will carry off floodwater faster, and in the process they destroy all the wild riparian edges that express a rationality that is different from economics, one we have not yet fully understood. The elementary need in learning how to farm water effectively, Leopold would have said, is to stop thinking about the problem exclusively as economists and engineers and to begin learning the logic of the river.

The oldest river in the world in terms of sustained human use is the Nile. Any modern farmer or agricultural expert would be well advised to study the natural history of that usage from its earliest times (about 5000 B.C.) to the present.[4] It is a history with good and bad lessons: an extraordinarily long symbiosis between people and river, a recent set of cataclysmic environmental changes, a persistent tendency toward concentrated political power based on water management, a rich accumulation of river lore given to Egyptian life. Beginning in the time of the pharaoh Menes (about 3200 B.C.), the valley farmers constructed a series of ditches and dikes to direct the annual floods to their fields. For thousands of years thereafter, Mother Nile fed the Egyptians with undiminished abundance, as the water and silt from the floods enriched the narrow strip of green land that lies between the river and the desert. When the river was low for lack of highland rains, there were poor harvests and famines. Eventually the Egyptians learned how to store the surplus of barley and pulse from the better years in order to feed themselves in the leaner ones. It was not dumb luck that gave their agriculture a durability unmatched anywhere else; they respected the river, learned to use it without violating its order, and thereby achieved an advanced level of civilization.

The Egyptians developed a technically simple basin system of irrigation that had little adverse environmental impact and was manageable under most circumstances by the ordinary fellaheen in the villages. Even so, historical evidence indicates that now and then local management proved inadequate. Government experts thereupon entered the scene, and on the foundation of their more unified, sophisticated program of water control, Egypt created a powerful, despotic state, which exercised life-and-death authority over the masses. More intensive, centralized control over the river became a means for some men to dominate others, giving rise to what Karl Wittfogel has called a "hydraulic society."[5] Egypt is one of history's outstanding examples of that socioecological order in which a concentrated power structure emerges out of large-scale water engineering and coordinated irrigation.

In the early nineteenth century, Egypt was invaded by British and French armies, whose engineers sought, as agents of empires, to convert that country to Western-style agriculture. Egyptians could raise more crops, they argued, and sell them at a good profit in the world marketplace, if dams and storage reservoirs were constructed along the river. Then began a long process of environmental transformation that, most recently, has led to the High Aswan Dam, a gargantuan chunk of concrete that has created Lake Nasser, one of the largest artificial bodies of water in the world. Now the silt that for so long fertilized Egypt's fields accumulates behind the dam. Downstream, new irrigation canals have stimulated an epidemic of snails, which carry the disease schistosomiasis into agricultural settlements, while the once productive delta fisheries are disappearing. In place of the old pharaonic hierarchy, the nation has substituted a powerful modern bureaucracy, which alone has the trained competence to design and maintain the hydraulic apparatus. What will be the life span of this new water regime? Not, assuredly, so long as that of the old one. Egypt has put economic calculation in the place of ecological rationality, short-run maximum production in the place of sustainability and the round river. The result may be a long-lasting decline of the Nile valley as an agricultural resource.

Americans have followed much the same course with rivers in our arid lands. Over the past century virtually every major western river has been dammed, diverted, and siphoned off to distant places, until the natural drainage of water has been obliterated over large parts. (A good many eastern rivers have also been altered, especially those under the Tennessee Valley Authority's administration.) The Bureau of Reclamation reported in the seventy-fifth year of its program a triumphant record of watershed manipulation: 322 storage reservoirs and 345 diversion dams, along with over 14,000 miles of canals, 35,000 miles of laterals, 930 miles of pipelines, 218 miles of tunnels, and more than 15,000 miles of drains.[6] The Missouri River project has been the most expensive, the Columbia scheme boasts the most massive structures, and the Central Valley of California turns the biggest agricultural profit. But it is along the Colorado River that the most celebrated showpieces of American water engineering have appeared; from the dedication of Hoover Dam in 1935 until the early 1960s, so much of the river was stored or sent through aqueducts to Los Angeles and the Imperial Valley that its water no longer reached the sea.[7] That was only the first phase in the bureau's ambition to achieve, in its own words, "total control for greater wealth." The Colorado has been regarded by a succession of planners as an unruly, dangerous beast that must be tamed and disciplined by modern science. Now it meekly furnishes water for man-made surfing playgrounds and citrus groves in Arizona, casinos and massage parlors in Nevada, manufacturers and movie studios in California, until it may be said that the only truly chaotic force left is the unrestrained human id, using the most advanced hydraulic techniques to waste a river on its most ephemeral whims and appetites.

One of the more substantial achievements of river control, though ultimately it too has become as irrational as desert surfing, has been the industrialization of American agriculture. Wherever intensive, large-scale irrigation has appeared, farming has quickly become a factory operation, mass producing for a mass-consuming market. Since at least the 1930s, the irrigated farms of the Southwest and the West Coast have led the nation in adopting the industrial mode, and they have forced farmers elsewhere to keep pace or drop out. Irrigation farming is expensive, requiring large amounts of capital investment; where there is no subsidy, only a small number of farmers can afford it. Once agriculture has started down that industrial road, it is not easy to stop: waterworks are followed by pesticides, chemical fertilizers, armies of stoop-and-pick laborers, and a high degree of mechanization.[8] The western river thus ends up becoming an assembly line, rolling unceasingly toward the goal of unlimited production. After basic human needs have been satisfied, there appears to be no deeply considered purpose justifying this production; water comes to have merely an international-trade value, abstracted from its natural milieu and made to serve the industrial imperative of growth as a self-justifying end. When we can "drink" no more irrigated oranges or corn or rice, world marketers tell us, we can sell our rivers (sell, that is, the products they water) to Japan or Germany. In that outcome is the final alienation of a people from their land and its stream of life—when both are sold to the uttermost parts of the planet for a mess of gadgets.

Alienation is an abstract, though completely real, outcome. Thirst is more concrete and measurable, and it is staring us in the face. The irrigated factory farms of the West are likely to drink the region dry. Irrigated crops currently use about one-half of this country's annual withdrawal of water.[9] But in western states with low rainfall that proportion is much higher: 80 or 90 percent. According to the U.S. Geological Survey, California withdrew in 1970 some thirty-three billion gallons of water per day for irrigation, or one-fourth of the national total, from surface and groundwater sources. Idaho was the second largest agricultural user, with a daily withdrawal of fifteen billion gallons; Texas came next, with ten billion gallons. Thirteen other states—all but one of them (Florida) located west of the Mississippi—used at least one billion gallons a day for irrigation.[10] In some places most of the water comes not directly from rivers but from long-accumulated underground deposits. Each year farmers pump from the Ogallala aquifer of the Great Plains more than the entire flow of the Colorado River. That resource, left over from Pleistocene times, once the largest natural storage system of its kind anywhere, now has a life expectancy of about forty years.[11] Irrigated farming, carried on in so grand a fashion, has become an extravagance this nation cannot afford and which many states cannot much longer sustain.

Even if there were enough water to last forever, in many cases there might not be enough energy to make it available. Modern irrigation involves the drastic reorganization of the hydrological cycle, and that task can succeed only with plenty of cheap energy. In the United States, it has taken an abundance of artificial energy to make our rivers move in unnatural ways, in ways that are less efficient in terms of their own dynamics. For in nature, Robert Curry explains, a river constantly seeks the most energy-efficient path to the ocean.[12] Wherever an obstacle appears, the river goes to work to remove it or find another route. Put a dam across a canyon and the river there immediately gets busy at washing it away. Somewhere humans must find a ready source of energy to keep that river blocked, to force it out of its bed and over tablelands and floodplains, or to lift it across mountain ranges to run in city pipes. Exhaust or lose that external source of energy to apply against the river and humans lose the ability to overcome the natural laws of watershed energetics. They must then let the water flow where it finds the going easiest. That is the prospect we are now facing in our man-made water regime.

In the ancient Egyptian world, the energy for water manipulation came from corvees, immense legions of peasants drafted by the government to build and maintain works, impelled by the whip when they got tired. The modern approach has been to substitute fossil fuels for much of that sweat. We have celebrated the change with expansive rhetoric: "unlimited abundance" and "plenty of water and electricity at the throwing of a switch." But no one has yet told us precisely and comprehensively how much energy it has required to erect works like Hoover Dam, to keep them in repair, and to pump their stored water away, nor have we been told how that energy demand compares with the hydropower they generate.[13] James Bethal and Martin Massengale have calculated that irrigation pumping and distribution alone consume 13 percent of all energy used in American agriculture. In a state like Nebraska, where the center-pivot sprinklers spread underground water over round, checker-like cornfields, ten times as much fossil fuel goes into irrigation as goes into all nonfarm requirements.[14] The mounting cost of fuel today may put the farmer out of that enterprise long before his well runs dry. Water cannot run uphill unless there is enough money to push it. Only the foolhardy will state unequivocally that no new source of low-cost fuel will ever be found, but it will be a bigger fool who will tell us that such a breakthrough can come with no strings attached, no undesirable side effects, no need to confront ecological limits. So long as we do not think as rivers do, our irrigated agriculture will always be an exercise in the ultimate futility of trying to repeal the natural laws of flow.

The decreasing supplies of water and energy are only the most obvious threats to the American irrigation empire. Perhaps a more serious, long-range nemesis is the salt poisoning of arable land, which seems to be an inevitable consequence of desert irrigation.[15] This is the problem of soil and water quality degraded through overuse. In regions of scarce rainfall, the earth contains a large amount of unleached salts; pouring water onto fields there brings those salts to the surface and into the river system. Continual stream diversions lead inexorably to poisoning downstream, for as the irrigation water evaporates from reservoirs or transpires from rows of plants, it leaves a whitish residue of salt behind. This salinization put the Mesopotamian irrigators out of business thousands of years ago. Today more than one-third of the world's irrigated land has salt-pollution problems that diminish the productivity of the soil and, in extreme cases, ruin it forever. There is very serious salinization of farmland in California, Hawaii, along the Rio Grande, and throughout the Colorado River Basin. The Coachella Valley near Palm Springs must use much of its canal water not to water crops but to flush away salt left behind by earlier irrigation. The nearby Imperial Irrigation District has already spent millions to keep ahead of creeping salinity and now hopes that taxpayers will foot the bill to stave off this self-induced destruction. No matter who pays for the remedies, the only cure is more water consumption, more drains to get rid of excess water quickly, more energy and capital for desalting installations— a cure that becomes at some point even worse than the illness. Is it really worth the risk of irreversible poisoning of the land to keep agricultural exports high? To have lettuce in January?

The list of the environmental problems caused by western irrigation schemes has grown longer and longer in the science journals. Where a considerable amount of groundwater is pumped out, the land may subside and form a great concavity, destroying roads, houses, and bridges, and disrupting the surface life.[16] When a river ceases to bring fresh water to the ocean, disaster strikes the biologically rich estuaries along the coast. Wastes put into the diminished current upstream can no longer be adequately diluted; oxygen in the rivers consequently gets depleted. In streams that have been dammed, temperature changes occur, killing the native fish if oxygen depletion does not. And only recently we have begun to investigate the impact that widespread irrigation has on regional climate patterns: irrigation water lost to evaporation, for example, may add significantly to the rainfall downwind.[17] In the face of such potentially destructive, always unpredictable possibilities, it has become clear that making "the desert blossom as the rose" is a far more complicated job than we once, in our juvenile innocence and self-assurance, believed.[18]

Now too we are beginning to admit what some critics have long claimed: that irrigation development in the American West has done much damage to a thriving agriculture and rural life in other regions. In congressional debate over the 1902 National Reclamation Act, eastern opponents protested that they were being taxed to create competition for themselves.[19] With a long sun-filled growing season, a national transportation network, and publicly subsidized water, irrigated farmers in the West did in fact come to enjoy a clear market advantage. Today the effects of the competition have been painstakingly studied and calculated; no longer can they be denied by western apologists. The Bureau of Reclamation's projects alone have replaced five to eighteen million cultivated acres elsewhere. The South, for instance, saw a sharp decline from 1944 to 1964 in cropland harvested—about one-third of its total. In the same period, western reclamation projects added more than 60 percent to their acreage, and much of it has been growing cotton the South once would have grown. Potatoes, wheat, feed grains, sugar beets, fruit, and vegetables have all shifted westward, too. If that were not enough, the federal government paid out in the same years as much as $179 million annually in crop-reduction payments to farmers on reclamation projects. What was added to cropland with one hand was taken away by the other, although in both cases eastern farmers did some of the paying, sending their tax money to the more arid region, then suffering from the low commodity prices brought about by the resulting overproduction, then shelling out reduction incentives.[20]

Even in the profit-maximizing terms of economists, there is little sense in American irrigation policy. It has made paupers of many living on the land, depleted farm communities, and sent the uprooted and defeated off to city unemployment offices. Those easterners leaving the farm have packed along in their suitcases a hard-acquired experience of working the earth.

Someday when we need again a fund of practical experience to till eastern lands, much of it may have been lost. Instead of investing in the preservation of existing know-how at the eastern grassroots, we have poured billions of dollars into the conquest of desert rivers, into technological virtuosity, into powerful government bureaucracies and agribusinesses. The result is an agriculture seemingly immune to the vagaries of climate, but in reality highly fragile as all leviathan systems are.

The West provides the most glaring evidence of the economic confusion and ecological irrationality of American thinking about water. But in the rain-rich states, too, there is ample testimony that we have not learned to farm the hydrological flow with consistent insight and prudence. Why else are we losing more soil to erosion than ever before? With all of agronomic and biological science at our fingertips, with a collective wealth beyond anything previously experienced, we allow five tons of topsoil to wash away from an average acre on an average farm every year. In western Tennessee not far from the Mississippi River, where landowners have plowed up their hilly pastures to plant soybeans, the annual soil losses average 30 to 40 tons per acre, and some farms lose 150 tons. Tennessee farmers, in the words of the Des Moines farm journalist James Risser, "have been caught up in the runaway growth of an American food-production machine and an inflation spiral in prices and may be mortgaging the future for quick profits today."[21] Old Man River says to those men and women: Keep the earth covered or I will take it away. But they cannot hear that warning so long as the louder roaring in their ears says: Make money—pay for that new tractor—produce produce produce.

It is time we began to rethink our agricultural relation to water. The problems are so numerous and complex that a flurry of quick-fix solutions is no longer adequate. What is needed is a fundamentally new approach to the challenge of how to extract a farm living from the hydrological cycle, both in humid and in arid regions. That requires vision more than technique: a way of perceiving, a set of mental images, an ethic controlling agricultural policy and practice. It demands, as I have said before, learning to think like a river.

For a long while this country has been perfecting a strategy of comprehensive river planning. We have called this approach "conservation," though it has drastically remade rather than conserved what nature has given us. It has always been based on technological instead of ecological thinking; planners have defined their task as taking entire river systems apart, putting them back together in more "useful" arrangements, and delivering the goods wholesale to the farmer.[22] A more reasonable strategy would be to focus on the individual farm, asking what its particular needs are and how they can be met with the least possible interference with the water cycle. Start with the local and specific rather than the general and grand. Develop a well-considered set of ends for public water policy. Seek then to meet those ends with the most elegantly simple means of water use possible, means that are economical, appropriately scaled to place and need, and capable of enduring indefinitely.

No farmer needs to grow corn in the desert or to use the Colorado River to do so. He may be inclined to choose to raise corn because he grew up in Indiana where corn was the traditional crop and then took his habits west, or because a county agent has told him that the world price of corn will be soaring after harvest. But that is not an analysis of need; it is a selection of methods. What the farmer really needs is a comfortable living for himself and his family, along with a chance to use intelligence and initiative to gain a measure of satisfaction from his work. The American consumer needs something to eat—nutrition and taste on the table at a not outlandish price. Can those genuine needs be met without turning the Colorado or Snake River into an elaborate artificial plumbing system, plagued by mounting costs and environmental destruction at every joint and faucet? Of course they can, if we are willing to put our ingenuity to work inventing, disciplining, and adapting; they can be met, that is, if we will undertake a radical reconstruction of methods.

The first specific step toward a new water consciousness is to end all federal subsidies of irrigation projects in the West.[23] The subsidies should not be halted abruptly, but gradually, reversing with care and sensitivity the existing policy that has been in effect nearly a century now. Americans have no reason to fear such a change. The greatest portion of artificially watered acreage in the West raises crops that can be grown more cheaply elsewhere: 37 percent of all federal reclamation land, for example, is used for hay and forage; 21 percent for corn, barley, and wheat; 10 percent for cotton.[24] The United States will hardly starve if we do not subsidize those crops, for farmers in the East will raise them instead, and they can do so in ways far less disturbing ecologically. But since western farmers have long been induced by government incentives to move west and set up their irrigation plans, they should not be made to suffer by this policy reversal. What is needed is a new "homestead program," equivalent to the one devised in the mid-nineteenth century, that will encourage many western farmers to relocate in the more humid areas and learn the best practices for those places. For most of our national history we have assumed that to go forward was to go west. Now a sustainable agriculture requires a redirection of progress: Go east, young man or woman, and grow up with the country.

The West is now overpopulated, grossly exceeding its natural river capacity, and a new sense of water limits should stimulate the region's city residents as well as farmers to resettle eastward where they can be supported more easily. Those who remain, who constitute a permanent population in equilibrium with the environment, must have new water technologies that will enable them to survive, enjoy a modest prosperity, and grow some of their food close at hand. Unfortunately, almost no official thought has been devoted to alternative technologies that can provide for that population, although we know that at some point the great reservoirs will be filled with silt. It is time, perhaps past time, to begin the process of reinventing the West. The farsighted desert or plains farmer will start now to work out his own salvation, not wait for the planners, although he can use the advice of hydrologists, geneticists, and engineers. He will study the art of adaptive dry farming. He will demand some new crop varieties that can survive in places of little rain, and perhaps he will convince legislators to grant some aid to ease that changeover, much as they have subsidized home solar-energy conversions. With his neighbors, he will devise ways of diverting floodwaters without appropriating the entire river, ways of guaranteeing a minimum flow in the channel to support its ecology while making use of the river for crops. Where local markets require fresh fruits, vegetables, and milk, he will install drip irrigation, which uses far less water than furrow methods. Confronting the decline of fossil fuels, he will let the sun and gravity do most of the work of lifting and circulating water through his farm. These are a few of the ways in which agriculture can begin to adjust to an arid setting. Rather than insisting that drylands be made over into a version of Missouri, producing crops for which they are ill suited, we should begin imagining a future West that is finely tuned to its unique environment.

In more humid areas, farmers face the challenge of not letting a wealth of rain wash away their common sense and with it their soil. The first principle of good water management, it still bears repeating, is the maintenance of soil cover. In many places this means restoring the natural forest vegetation or planting the strong defense of a tree crop. In other places a thick sponge of grass will be enough to absorb the impact of falling water, slow its race to the sea, and keep the rivers clean and sweet. New perennial crops that make plowing unnecessary can cut erosion losses to natural replacement levels. Diminished use of pesticides and chemicals can reduce groundwater contamination. These familiar therapies are all parts of a larger vision: an agriculture in which every farm fits harmoniously within the dynamics of its own local watershed, rather than an agriculture in which every farm seeks to maximize its share of the money economy.

By now it should be evident that no market will ever pay farmers for accommodating themselves to their watershed. To be sure, the market-place will reward long-range calculation more handsomely than many farmers are aware. But finally the marketplace is an institution that teaches self-advancement, private acquisition, and the domination of nature. Its way of thinking is incompatible with the round river. Ecological harmony is a nonmarket value that takes a collective will to achieve. It requires that farmers living along a stream cooperate to preserve it and to pass a fertile world down to another generation. It requires that urban consumers be willing to pay farmers to use good conservation techniques as well as to produce food. Without a public willingness to bid against market pressures, there will not be a radical reconstruction of farming methods or a rapprochement between agriculture and nature.

Americans, like people in other places and times, have a history of considerable violence toward the land and its life-giving rivers. Perhaps we have done more violence than most nations—certainly we have done more damage in a shorter period than most. Violence is typically a sporadic act, ill considered and destructive to the perpetrator as well as the victim; it is never the basis for permanence. What is now required in our agriculture, if it is to be secure, is a rejection of violence. Fortunately, we still can find in this country a broad enough margin of resources that we are not forced to violent land and water use; we can make other choices and avoid frantic, draconian measures.[25] We are in a position to think not only about self-preservation but also about generosity and peace—about ethics.

Almost forty years ago, Aldo Leopold wrote that we will never get along well with nature until we learn to regard it morally. We must develop, he maintained, a sense of belonging to the larger community of nature, a community that has many interests and claims besides our own. We must cultivate a moral sensitivity to that community's integrity and beauty. He spoke of the need for a "land ethic," including in it a moral responsiveness to all parts of the ecological whole.[26] But given the centrality of water in our lives, and given the magnitude of the problems we confront in farming our watersheds, it also makes sense to talk about a "water ethic." Water, after all, covers most of this planet's surface. Even more than land, water is the essence and the context of life, the sphere of our being and that of other creatures. It has a value that extends beyond the economic use we make of it on our farms. Preserving that value of water through a new American agriculture is an extension of ethics as well as of wisdom.

6

Energy and Agriculture

Amory B. Lovins
L. Hunter Lovins
Marty Bender

American agriculture both uses and produces energy. "Modern" farming pours fossil fuels into the soil and mines food energy in return. But the balance is not even. The great combines, the center-pivot irrigators, the food processors, the trucks that haul the average molecule of U.S. food 1,300 miles before someone eats it, all use energy.

As a result, slightly more than two calories of energy are invested per calorie of food obtained for all agricultural production in the United States. If we look at just the agricultural production consumed within the United States, slightly more than three calories of energy are invested per calorie of food obtained. When the energy costs for processing, distribution, and preparation are added onto the three calories, the total energy cost is about 9.8 calories of energy per calorie of food consumed in the United States. Straightforward multiplication shows that the U.S. food system uses 11.5 quads of energy annually to feed 234 million Americans with a diet of 3,420 calories per person daily. (One quad is one quadrillion

The authors wish to thank Bryant M. Patten II and Laura Jackson for their good and hard work in preparing early drafts of this chapter.

British thermal units, or BTUs; total U.S. energy use in 1974 was 72.8 quads.) In contrast, the food systems of the rural populations of developing nations use an estimated 16.4 quads annually to feed about two billion people with a diet ranging from 1,800 to 2,400 calories per person daily (such diets contain much less meat than ours). This 16.4 quads is less than one-tenth of what the same number of people would consume were they utilizing the food system of the United States.[1]

This comparison of energy use in food systems may be meaningful for analyzing short-term aspects of agriculture, but it tells us nothing about the agricultural degradation of soil and water resources in these systems. The sustainability of an agriculture depends not only on the balance between the energy it uses and produces but also on its ability to preserve its soil and water resources. In many localities throughout the world, primitive and modern agriculture alike have badly eroded the soil and ruined the quality of the water.

In the United States, during the past 200 years, at least one-third of our cropland topsoil has been lost. Studies within the past decade indicate that we are losing from 25 to 50 percent more soil per acre now than we were in the 1930s when Hugh Hammond Bennett, founding chief of the Soil Conservation Service, was lamenting the loss of American soils.[2]

Since that time, water has also become a serious resource problem. A full two-thirds of the groundwater pumped in the United States is used to irrigate our crops. About one-fourth of this withdrawal is overdraft (water drawn at a greater rate than it refills).[3] Our agricultural system seems to have a penchant for "double mining": we now mine fossil fuel in order to mine water; long ago we became accustomed to mining soil nutrients during harvest while promoting the downward rush of the remaining soil to the sea. Conserving these natural resources is at least as vital to our interests as reducing the massive amount of fossil energy that American agriculture requires.

But agriculture may face even greater strains than its own energy imbalance. The oil shortage encouraged many people to look to agriculture to provide substitute fuels. Alcohol from biomass offered the hope of a renewable domestic source of liquid fuel to provide the mobility on which modern society depends. A major biomass fuels program, however, would hold a great risk: if it causes us to regard fuels as more important than soil and water, then it would only contribute to the collapse it was meant to help forestall. Thus, we urgently need to understand the present role of energy in agriculture and to seek sustainable ways to provide both food and fuel for the future.

PETROFARMING

America has become the shining example of agricultural and technological advance. Our farms have provided an ever-higher yield from the land while requiring fewer and fewer people to work on the land. Supermarkets offer a staggering array of produce, available almost regardless of season or weather. We produce this abundance in such quantities that our food exports have earned us a reputation as the breadbasket of the world.

What is less well known is that the past forty years of more food and fewer farmers were possible only because of a temporary overabundance of subsidized fossil fuels. Although thoughtful analysts have for years pointed out that highly mechanized, chemicalized, and capitalized farming cannot be sustained, cheap energy has so far enabled the agricultural sector to cover most of its shortcomings. However, as energy prices rose in the late 1970s, the era of cheap energy ended and so did the conventional idea of what farming might look like in the future. The dependence of U.S. agriculture on fossil fuels was probably not so obvious to the American public because agriculture consumes only a small part of the total U.S. energy budget (Tables 1 and 2) and because the increases in agricultural energy use during the past forty years were gradual. As of 1974, on-farm energy use was 2.9 percent of the total U.S. energy budget, and food processing, distribution, and preparation accounted for 13.6 percent. The American public

Sector	Energy Consumed (quadrillion BTUs)	Percent of Total
Agriculture (on farm)	1.46	2.0
Mining	2.04	2.8
Construction	1.97	2.7
Manufacturing	21.00	28.8
Transportation	17.40	24.0
Commercial	5.70	7.8
Household	11.60	15.9
Electric utility	11.60	16.0
Total	72.77	100.0

TABLE 1. 1974 Energy Consumption by Sector
Source: Based on Wendy Kolmar, "The Energy Requirements of U.S. Agriculture," *Monthly Energy Review*, Department of Energy/Energy Information Agency-0035 (83/2), p. 1.

Sector of Food System	Energy Consumed (quadrillion BTUs)	Percent of U.S. Total Energy Use	Percent of U.S. Food System Energy Use
On farm			
Direct energy	1.25	1.7	10
Indirect energy	1.02	1.4	8
Food processing	3.6	4.9	30
Distribution system	1.2	1.7	10
Commercial food service	2.0	2.7	17
Home food preparation	3.1	4.2	25
Total	12.17	16.6	100

TABLE 2. Energy Consumption within the U.S. Food System

Note: There is a minor discrepancy between the on-farm numbers listed here and in Tables 1 and 3 because they were derived from different references.

Source: Based on Booz-Allen and Hamilton, Inc., *Energy Use in the Food System*, Office of Industrial Programs, Federal Energy Administration, FEA/D–76/083 (Washington, D.C.: U.S. Government Printing Office, 1976); U.S. Department of Agriculture, *Energy and U.S. Agriculture: 1974 and 1978*, Economics, Statistics, and Cooperatives Service Statistical Bulletin no. 632 (Washington, D.C.: U.S. Government Printing Office, 1979), p. 64; John S. Steinhart and Carol F. Steinhart, "Energy Use in the U.S. Food System," *Science* 184, no. 4134 (1974): 33–42; and R.A. Friedrich, *Energy Conservation for American Agriculture* (Cambridge, Mass.: Ballinger, 1978), pp. 55, 69, 104.

also does not see the energy that is embodied in the manufacture of fertilizers and pesticides, in the packaging and transportation of food, etcetera. For example, fertilizers account for about 30 percent of U.S. on-farm energy use, a fact unknown to many farmers (Table 3).

Economic factors, along with scientific knowledge and technology, virtually forced the extensive use of energy to increase crop yields and to reduce labor in order for individual farmers to survive economically in American agriculture. Following the Dust Bowl of the 1930s, midwestern farmers sought desperately to increase soil productivity. After World War II, that search happened to coincide with a glut of refined products from the oil industry, which had scaled up to fuel a nation at war. For what appeared at the time to benefit both, agriculture began to use the products of the oil industry.

The early uses of oil-derived fertilizers and pesticides yielded remarkable results: the profits made from the use of deceptively inexpensive fossil-

Component	Energy Consumed (quadrillion BTUs)	Percent of Total
Direct		
Field operations	0.46	20
Transportation	0.26	12
Irrigation	0.25	11
Livestock, dairy, and poultry	0.16	7
Crop drying	0.12	5
Subtotal	1.25	55
Indirect		
Fertilizer	0.65	29
Equipment manufacture	0.30	13
Pesticides	0.07	3
Subtotal	1.02	45
Total	2.27	100

TABLE 3. U.S. On-Farm Energy Use

Source: Based on USDA, *Energy and U.S. Agriculture: 1974 and 1978*; Steinhart and Steinhart, "Energy Use"; and Friedrich, *Energy Conservation*.

fuel feedstocks were irresistible. Fossil fuels allowed the farmer to overwork the land while still increasing the yield. As this soil mining accelerated erosion and diminished soil quality, more and more fertilizer was needed to maintain crop production. From a basis of twelve inches of topsoil or less, corn yields are reduced annually by an average of about 4 bushels per acre for each inch of topsoil lost; oat yields are reduced an average of about 2.4 bushels per acre; wheat yields are reduced by an average of 1.6 bushels per acre; and soybean yields are reduced by 2.6 bushels per acre. The primary reasons for the reduced yields on eroded soils are low nitrogen content, impaired soil structure, deficient organic matter, and reduced availability of moisture. So, as soil quality is diminished, more and more fertilizer is needed to maintain crop production.[4]

Pesticides showed similarly diminishing returns. Although the chemicals were at first highly successful, many insects have quickly evolved resistance to insecticides, and similarly, a number of weeds have evolved resistance to some herbicides. Instances are now legion in which pesticides such as DDT and parathion unfavorably depressed predator and parasite populations, thus allowing insect pests to multiply so that farmers had to use even more pesticides. Around 1948, at the outset of the synthetic insecticide era, when the U.S. used 50 million pounds of insecticides, the insects destroyed 7 percent of preharvest crops. Today, under a load of 600 million pounds, we lose 13 percent of crops before harvest.[5]

A third agricultural use of petroleum is as fuel for farm machinery. As the economics of the "get big or get out" syndrome forced farms to become more mechanized, farmers were forced to manage larger and larger acreages in order to pay off the debts on equipment. So tractors, implements, and irrigation systems grew in scale to meet the greater demands and consumed ever more fossil fuels.

The addiction to petrofarming came on so gradually and insidiously that it required serious price shocks, as happened in 1974 and 1979, to bring this problem to public attention. Today, American agriculture uses more petroleum products than any other industry in the nation.[6] The largest direct agricultural user of energy is farm machinery (this includes trucks and automobiles), which accounts for almost three-fifths of all directly used energy. In 1978, gasoline and diesel fuel used for all purposes on U.S. farms totaled 0.72 quads nationally. Energy to dry crops to prevent spoilage amounted to a further 0.12 quads. And energy to maintain livestock, dairy, and poultry totaled 0.16 quads.

Another major direct use of energy is irrigation. In 1975, to irrigate just the forty-eight million agricultural acres in the seventeen western states (this accounts for 85 percent of all irrigation in the nation) required one-quarter of one quad and this high energy use makes itself worse. Pumping up groundwater from the San Joaquin and Imperial valleys, for example, has made the water table recede so that increasing amounts of electricity are needed to pump it; agriculture is California's biggest single user of electricity. In the Pacific Northwest, part of the justification for building several nuclear power plants under the Washington Public Power Supply System was for irrigation of the western slope desert of Washington state. WPPSS is now sliding toward the largest municipal bankruptcy in history.

The energy embodied in chemicals and equipment—indirect energy use—is not much less than direct use. About forty million tons of fertilizer are applied to America's fields each year—approximately 330 pounds for each person in the country. The feedstock, manufacture, and transportation of fertilizers consumed 0.65 quads of energy in 1978. Similarly, pesticides are "a fossil-fuel based system of control, for a full 80 percent of the one billion pounds spread annually comes out of oil wells." Pesticides required about 0.07 quads of energy in 1978 for feedstock, manufacture, and transportation. The energy required to purify pesticide-contaminated water supplies should also be included. The energy embodied in manufacturing equipment is estimated to be 0.3 quads.[7]

Farmers continue to depend on fossil fuels for four basic reasons. First, these methods, gradually adopted over the past four decades, are capital intensive. A farmer who has invested in a large combine, a center-pivot irrigation system, or an expensive pesticide applicator is understandably reluctant to change to a relatively untested technology that might require new techniques or equipment. The farmer may also be too tied up in loan payments to have the financial flexibility to change his or her operation.

Second, there are long time lags in perceiving that a process is yielding diminishing returns—that greater amounts of pesticides are protecting less fruit. If a process has worked better each year for a long time, the farmer will be slow to see that it has begun to work poorly or not at all.

Third, in addition to the inertia in the farmer's mind, there is enormous pressure from the agribusiness lobby, which has a vested interest in keeping American agriculture hooked on oil. Farm magazines prosper because of advertisements from chemical, tractor, and irrigation companies. Such companies also are able to dribble money into the universities for research grants on their products. They thereby gain up to a fivefold leverage on public money because university scientists have more personal control over private grant funds than over public money, which is used mostly for maintaining research facilities. The cash cost to these companies is minimal. Thus, virtually all the sources of information to which a farmer might turn reinforce dependence on fossil fuels.[8]

Fourth, petroleum-powered machines were substituted for human and animal labor eliminated by the substitution, thereby making it difficult for farmers to switch from machines back to human and animal labor. In 1920, there were 6,448,000 farms in the United States. The farm population at that time was 30 percent of the U.S. population, and there were twenty-five million horses and mules. By 1978, the number of farms had dropped to 2,436,250, and only 3 percent of the U.S. population lived on them. The number of horses and mules had dropped to about two million around 1962, and though it had risen to about eight million by 1978, almost none of these animals are used for farming.[9]

Both ecology and economics are conspiring to put an end to petrofarming. The inflation caused by the escalating price of fossil fuel is giving farmers little choice. Either they find new fuel-efficient farming methods or they collapse under a tremendous debt burden. Farm debt has doubled in the past six years; in 1982, it was $216.3 billion. Debt now represents about $10 per dollar of annual farm net income—about 4.4 times the 1973 ratio.

This implies that the average farm would lose money at annual interest rates above about 10 percent even if it had zero operating costs.[10]

ENERGY FOR FOOD PROCESSING

On-farm energy use accounts for only 18 percent of the energy consumption within the U.S. food system. Food processing and distribution account for 40 percent, and preparation accounts for the remaining 42 percent (Table 2). Since consumer demands dictate how food is processed, distributed, and prepared, we should look at what happens beyond the farm.

The food-processing industry has become an important support industry for the farmer. In the past fifty years, canned, frozen, and other processed foods have become the principal items of our diet. During the 1960s, there was a slow but steady shift toward consumer consumption of more energy-intensive food, such as beef and highly processed foods. More than three-quarters of the food grown on farms is processed before shipment to the consumer. The food-processing industry is the fourth largest energy consumer of the Standard Industrial Classification groupings, with only primary metals, chemicals, and petroleum ahead of it. In 1974, food processing and distribution accounted for 4.8 quads or 6.6 percent of total U.S. energy use (Table 2).[11]

The eleven most energy-intensive food-processing industries as a whole derive about 48 percent of their energy from natural gas, 28 percent from electricity, 9 percent from coal, 7 percent from residual fuel oils, and 8 percent from other fuels. Most processing plants in these industries were designed and built during an era of cheap energy and thus are energy inefficient. Energy losses to the environment (such as waste heat from buildings and processing equipment and hot water) decrease the useful amount of energy available for processing. For example, only 34 percent of the energy put into vegetable canneries in western New York actually goes into food processing.[12]

Finally, food-processing waste represents another energy loss. The energy cost of the processed food includes the energy invested in the processing and embedded in the waste. Additionally, as a consequence of the material losses, more raw food is required to obtain a given amount of processed food. The food-processing industries annually generate 14.4 million tons of solid waste, which is 9.6 percent of that generated by manufacturing industries in the United States every year. The 262 billion gallons of wastewater produced by the food-processing industry represent a twelve-day supply of water for U.S. urban domestic use.[13]

SEEDS OF CHANGE

Since 1930, many farmers have paid the carrying charge on their debt by borrowing against inflation in land values. When land prices stabilized in 1980, farm foreclosures spread across the Midwest at a rate unseen since the Great Depression. Despite an intensive export drive, so many farmers are now unable to carry the debt they incurred for heavy machinery and irrigation equipment that their bankruptcies in turn are threatening the solvency of major banks. In 1982, farm production expenses of $140.1 billion had almost caught up with farm cash receipts from crops and live-stock of $144.6 billion. In 1983, as this is being written, government farm price supports are estimated to be $21.2 billion, which is almost as much as the 1982 net farm income of $22.1 billion. With the Payment-In-Kind program (one form of crop-reduction payments) and the summer drought, government subsidies will probably exceed net farm income in 1983, for the first time in history. Yet most farmers can only go deeper into debt be-cause they are advised only on how to raise production—not on how to cut the costs of production in water, energy, chemicals, and machines.[14]

The ecological damage caused by petrofarming poses an even greater threat to Americans than does farm economics. The heavy use of nitrogen fertilizer has resulted in dangerously high nitrate levels in the groundwater of some areas. Drinking water contaminated by nitrogen can cause severe health problems. Pesticides are also contaminating groundwater in some areas. Heavy irrigation, made economically possible by cheap oil and gas, is eroding the soil and rapidly salinizing what is left. Soil erosion threatens farm productivity, and eroded soil is filling streams and rivers with silt and the air with dust.

Either the economic or the ecological failure alone should be enough to bring about a change in agricultural methods. Together, they make an urgent need for change. But such a change need not be disruptive or demand sacrifice, if it is done sensibly and soon enough. Like energy use in general, energy efficiency on the farm can be increased a great deal without reducing productivity.

For example, in 1976 the Center for Rural Affairs established the Small Farm Energy Project, a research and demonstration program in energy-saving technologies to increase the incomes from low-income grain and livestock farms in Cedar County, Nebraska. Cooperating farmers attended a series of educational sessions about efficient energy use and innovative technologies. The farmers themselves were responsible for the construction and monitoring of energy innovations. The project staff supplied design guidance and assistance. After three years, a total of $51,175 ($29,699 from the cooperating farmers and $21,476 in cost-sharing by the Small Farm Energy Project) was invested in the purchase of materials for 148 energy innovations. The farmers were saving about $27,312 per year and were using about 15 percent less energy than the control group with no difference in production.[15]

Conservation and efficiency improvements can range from weather-stripping buildings to changing the timing of irrigation systems to making better use of farm machinery. The last opportunity alone represents a saving of 0.1 quad or 21 percent over present use. Almost half of this figure could be saved solely by switching over from gasoline to diesel engines. Another potential area for immense energy saving is in irrigation, which in 1974 used some 260 trillion BTUs. Some experts report that half of the energy used in irrigation could be eliminated by improved techniques and by better pumping equipment. If farmers were to use low-temperature grain drying where climate permits, solar drying could save half the supplemental heat required.[16]

The energy consumed in fertilizer applications can in many cases be much reduced, either because present usage is more than is necessary or because at least part of the nutrient additives can come from organic, rather than inorganic, sources. Only one soil test is performed for every 162 acres planted, so the nutrient levels of most cropland are not known. For example, in 1974 in Illinois, 40 percent of corn and soybean fields were found to have greater than suggested levels for phosphorus, and more than 20 percent were higher than suggested in potassium.[17]

Livestock manure can be substituted for inorganic fertilizers with considerable energy savings. U.S. livestock manure production is estimated at 1.7 billion tons per year. More than half of it is produced in feedlots and confinement rearing. If one-fifth of the manure from confinement rearing and feedlots were used as fertilizer, it could serve seventeen million acres at ten tons per acre and save 0.07 quads. At the same time, manure would add organic matter to the soil, which increases beneficial bacteria and fungi, makes plowing easier, improves soil texture, and reduces erosion.

Rotating crops with legumes can supply nitrogen to cropland in considerable quantities. Although inorganic nitrogen is commonly added to cornfields at the rate of 112 pounds per acre, it is possible to plant legumes between corn rows in late August and plow them under as green manure in early spring. In the northeastern United States, corn and winter vetch planted in this way add 150 pounds of nitrogen per acre. If this procedure were performed on fifteen million acres of corn, which is about one-fourth of U.S. corn acreage, about 0.07 quads would be saved.

Increased efficiency in the manufacture and use of pesticides would conserve an estimated 0.01 quads per year. Using pesticides only where and when they are necessary would reduce pesticide consumption by up to one-half. This would amount to a saving of 0.03 quad, but the major benefit from decreased pesticide use is not the energy saved; rather it is the reduced contamination of the countryside.

Because of the soil erosion and energy costs associated with moldboard plowing, various forms of conservation tillage are being rapidly adopted, such as reduced-till, minimum-till, and no-till. Conservation tillage can reduce erosion on many soils from 50 to 90 percent. Additional benefits of conservation tillage are lower costs for equipment, labor, and fuel, and increased soil moisture retention. In 1982, more than 100 million acres—25 percent of all U.S. cropland—was farmed with conservation tillage practices. About 10.5 million acres of this were in no-till. By the year 2000, the U.S. Department of Agriculture estimates that 85 percent of all U.S. cropland will be in conservation tillage, of which 153 million acres will be in no-till.[18]

No-till can reduce fuel use in field operations by as much as 90 percent, and in some instances, it has actually increased crop yields. However, this energy saving is partly offset because no-till requires 30 to 50 percent more pesticide than conventional tillage needs. In addition, no-till soil sometimes needs extra nitrogen. The result, as numerous studies have shown, is that net energy savings for no-till on individual fields range from zero to about 10 percent; occasionally, savings run as high as 30 percent.[19]

Conservation tillage has inherent problems that include increased pest populations (insects, nematodes, rodents, fungi), increased susceptibility to plant disease, herbicide carry-over that locks farmers into continuous planting of corn, evolution of weed resistance to herbicides, and shifts in weed species, so that perennial weeds become a problem for which there are no fully effective chemical solutions. In addition, the farmer takes a greater economic risk. In management with no-till, farmers are not supported adequately by available technology or site-specific information services. If an herbicide fails to control weeds, the farmer usually cannot follow up with cultivation to salvage the crop.

Disconcertingly little is known about the effect of herbicides on nontarget organisms, including humans. Of the 150 chemicals used as herbicides throughout the world, complete metabolic pathways are known for only 3 or 4. Metabolic degradation of the remaining compounds is known only in part. For example, it has been found that atrazine (a principal herbicide used in conservation tillage) is chemically transformed into mutagenic substances by the corn plant and by conditions found in the human stomach. Thus, the present contamination of many midwestern drinking water supplies with atrazine is a cause for concern.[20]

Conservation tillage has become too reliant on herbicides to control weeds. Compared to research done on herbicides, little effort has been made to combine conservation tillage with integrated pest management, the system for managing insects, diseases, and weeds by combining resistant crop varieties, beneficial organisms, and crop rotations, plus other techniques. Such management uses pesticides only where necessary. One example of the research being done is a recent conference in the Mid-Atlantic region that focused on the use of crop rotation in no-till to break the cycles of weeds, insects, and diseases. Some farmers successfully use crop rotation with minimum-till without herbicides.[21]

Multiple cropping (intercropping or relay cropping) can also be used to control weeds in no-till. Again, most but not all of the research on multiple cropping involves the use of herbicides to kill or weaken the grass or legume sod (annual or perennial) several weeks before the corn is planted by no-till. Herbicides are also used in research to control weeds in the interplanting of grasses or legumes with corn.

The reduction of herbicide use in conservation tillage would bring larger net energy savings and less contamination of the countryside. Too often conservation tillage with herbicides is regarded as the solution to soil erosion problems. This viewpoint discourages funding of research needed to develop alternative methods of dealing with soil erosion.

A sustainable agriculture has to result in significant energy savings. It has been estimated that the direct energy savings from such agriculture could be 0.8 to 0.9 quads. Combined with the indirect savings of oil formerly used to replace soil fertility, the total would be 1.1 to 2.2 quads, or a financial savings of $8.5 to $17.4 billion (1980 dollars) each year.[22]

Clearly, it is imperative for farmers to implement techniques that remove them from the fossil-fuel treadmill. Increasing the efficiency of energy use and seeking new production techniques that minimize energy use are both necessary first steps. Ultimately, however, farmers—like all Americans—must find a replacement for fossil fuels. Thus, interest has turned to biomass fuels.

BIOMASS FUELS: PART OF THE SOLUTION OR PART OF THE PROBLEM?

Of all U.S. delivered energy, 58 percent is required in the form of heat, with three-fifths of the 58 percent at temperatures below the boiling point of water. A further 34 percent of our energy is needed in the form of liquid fuels to run our vehicles. Only 8 percent is needed as electricity. Heat is readily supplied by passive and active solar technologies that are commercially available already. Electricity, to which most of our national energy effort has traditionally gone, is already in oversupply: the power stations we have now can provide more than twice as much electricity as we need. Moreover, efficiently used electricity can be supplied just by existing hydropower, small-scale hydro, and a bit of windpower, without a need for any thermal power stations. For the production of both heat and electricity, the best renewable technologies now on the market actually cost less than new central power stations to deliver the same energy services. But this leaves unanswered the question of how best to provide liquid fuels for transportation.[23]

A renewable liquid fuel program based on biomass feedstocks must adhere to four principles if it is to be truly sustainable—not merely an alignment of soil mines:

1. *The land comes first.* All operations must be based on a concern for soil fertility and long-term environmental compatibility.

2. *Efficiency is vital.* Both the vehicle for which the fuel is intended and the means of converting the biomass into fuel must be efficient.

3. *Wastes are the source.* Use farming and forestry wastes as the principal feedstocks; no crop should be grown just to make fuels.

4. *Sustainability is a goal.* The program should be a vehicle for the reform of currently unsustainable farming and forestry practices.

PUT LAND MAINTENANCE FIRST

At present, much of American farming and forestry is little more than a mining operation. A massive biomass fuels program that simply serves to put greater pressure on overstressed land would not only risk crushing a budding energy program but could also pull down much of American agriculture. Renewable must mean sustainable in the very long run. No biomass program can long endure unless it is based on the preservation and enhancement of soil fertility, water quality, and the biotic community on which agriculture depends.

Not surprisingly, the biomass fuels program generally sanctioned and politically popular—a subsidized, corn-based ethanol/Gasohol program—generally ignores these concerns. While opponents and fans argue the net energy yields, the impact of such a program on the soil is overlooked. An alcohol fuels program must take into account not only the obvious and direct costs but also such hidden requirements as soil and water. Much of modern corn production is a real soil killer. The average corn farmer who never rotates crops loses around twenty tons of soil per acre with conventionally tilled corn. This is the equivalent of 2.3 bushels of soil lost per bushel of corn harvested. Corn grown west of the hundredth meridian, which passes just west of Hays, Kansas, is also energy intensive: energy for irrigation accounts for 66 percent of the energy used in growing irrigated corn in Kansas.[24]

Most criticisms leveled at the Gasohol program are really criticisms not of biomass fuels but of the agricultural system whose products are made into those fuels. Thus the first requirement of a sustainable biomass fuels program is to choose feedstocks whose production is not energy intensive and does not cause intolerable soil erosion and degradation.

DEVELOP EFFICIENT TECHNOLOGY

No kind of fuel supply makes economic or ecological sense without cars, trucks, buses, trains, ships, and aircraft that are efficient in fuel use. Current technology could give us an efficient vehicle fleet without sacrificing comfort or performance. The current U.S. car fleet is approaching seventeen miles per gallon, but the average import in 1979 was rated at thirty-two miles per gallon (mpg). A diesel Rabbit gets over forty mpg; a turbocharged Rabbit with the same performance as the diesel would get around sixty-four mpg. Volkswagen has already tested, presumably for more than academic interest, a 3,400-pound car, powered by a diesel engine and generator with a few batteries, which got eighty-three mpg; in 1981, Volkswagen tested an advanced diesel Rabbit that did eighty mpg in the city, 100 on the highway.[25]

The United States now uses approximately seventeen quads of liquid fuels annually for transportation. An efficient transportation system—using state-of-the-art trucks, buses, aircraft, trains, and ships—would use only five to six quads per year. Such an efficient vehicle fleet is not just a nice complement to a biomass fuels program; it is a prerequisite for making any liquid fuel supply affordable (and any biofuels program sustainable). The effort to supply seventeen quads of liquids per year has been costly enough in the oil era. Loading such a burden onto the already stressed agricultural base, even if technically possible, would yield most unpleasant dividends. However, a national priority of efficient vehicles, leading to a much reduced supply requirement, would bring our liquid fuel needs more within the capabilities of a sustainable biomass program. The turnover time for the U.S. car and light truck fleet is about a decade. For a cost of about $50 billion (1980 dollars), Detroit could be retooled to replace the present fleet with efficient vehicles.[26]

Almost as important as an efficient vehicle fleet are the proper scale of the biomass conversion techniques and their efficiency. Proper scale is essential to minimize the costs of collecting and transporting the feedstock. No definitive study has yet shown what scale is most cost effective, and there is probably no universal answer to this question, but the same diseconomies of large scale that have lately been described for electrical generating plants will probably dictate that biomass systems be relatively small in most circumstances.[27]

Biomass conversion plants could be as small as mobile pyrolyzers; these devices heat a feedstock—usually a woody one—with little or no ash, producing, among other possible products, char, a low-grade fuel and gas, and a heavy oil akin to condensed woodsmoke. They could be hauled on the back of a truck to go wherever there are small collections of feedstocks. A conversion plant the size of a milk-bottling plant could serve a half-dozen towns.

Initially, integrated systems that use both the fuel product and its by-products at the conversion site—the farm, the food-processing plant, or the pulp mill where the feedstock is available—make the most economic sense. For example, a pulp mill could utilize the steam from pulping processes to power a pyrolyzer plant to produce pyrolysis oil or methanol. The same steam could then be run from the pyrolysis plant through a turbine to generate electricity.

The goal in all of the considerations of scale should be to collect feedstocks with a minimum amount of transportation; the trace nutrients could be redistributed after the conversion. This transportation issue is especially important in the case of wet feedstocks, which tend to be heavy or perishable or both. Transportation can often be reduced by obtaining feedstocks from industrial plants where the wastes are already gathered into one location or, if necessary, by building the biomass conversion systems near the industrial plants. Where possible, the fuel should be sold locally, preferably through existing networks, again reducing distribution costs.

It is sometimes argued that many dispersed biomass fuel plants cannot produce enough total fuel to be nationally important. Even if this were true, a biofuels program meeting most of the needs of farms themselves would be important. Farms now get some 93 percent of their fuel from oil and gas and are often on the end of long and precarious supply lines. When world oil trade is disrupted, farmers suffer a double whammy: curtailed supplies boost fuel prices, while curtailed crop exports depress crop prices, making farmers even less able to compete for the scarce fuels.

A simple analogy, however, suggests that a large number of dispersed biomass fuel plants could deliver far more fuel than farms would need. Some 2.2 million dairy farms, with herds averaging fifty cows each, currently produce about 15 billion gallons of milk per year. That is one-fifth as many gallons as the gasoline used by American cars today, or about as many gallons as an efficient U.S. car fleet would use. Yet the milk is produced in relatively decentralized units spread around the country much more cheaply than if it all came, for example, from a few giant dairy farms in Texas and were pipelined around the country. Likewise, 21 percent of American oil comes from hundreds of thousands of scattered stripper wells that can be seen in farm fields; the average well lifts only about 2.8 barrels per day.[28]

Conversion efficiency will decide whether a biomass fuels program will result in a net gain or a net loss of energy. The conversion technology assumed in most official studies of biomass fuels is borrowed from other processes that are inappropriate for biomass conversion. It is relatively easy to show a net energy loss if one uses old, inefficient ethanol distillation technology, or thermochemical conversion processes designed for coal, which cooks more slowly, less efficiently, and at higher temperatures than biomass. But it is also easy to find more efficient methods. Ethanol stills, for example, have been, and can still be, enormously improved. A few years ago it took 50,000–100,000 BTUs to distill a gallon of ethanol to 190 proof. Today some commercial processes need only 25,000 BTUs to go all the way to 200-proof ethanol, and the best demonstrated processes have reduced this to a mere 8,000–10,000 BTUs. Stills can also use solar or, in some regions, geothermal heat. (For many applications, the alcohol also does not need to be completely dehydrated.)[29]

Distillation is a brute-force technology, but it is not the only option available. New England villagers and Appalachian moonshiners have long used freezing to fortify their applejack and corn squeezings. In most northern regions, this simple measure saves energy, although it does not provide anhydrous alcohol. Other alternatives for alcohol-water separation include hydrophobic plastics, chemical extractants, synthetic membranes, critical fluids, and cellulosic absorbants. More research is likely to yield attractive dividends with better technologies.

Finally, matching the feedstock, conversion process, fuel, and vehicle can greatly improve the efficiency of the whole system. Methanol, for example, offers exceptional performance in high-compression engines, especially in piston engines in aircraft at high altitudes. Methanol is currently cheaper (in general) than ethanol, especially when made from the woody feedstocks that are most commonly used. Methyl and ethyl esters, made by reacting methanol or ethanol with vegetable oils (preferably inedible ones) in a simple catalytic solar cooker, provide a superior diesel fuel. Some engines can do well with blends of ethanol, methanol, *tert*-butanol, and perhaps higher alcohols; others can use pyrolysates. Some engines can run cleanly and efficiently on natural, untreated vegetable oils. So many permutations are possible that nobody knows which will win in the end—only that there is a rich array of attractive possibilities.[30]

PUT WASTES TO USE

Most biomass studies assume the use of special crops, notably grains, grown specifically for conversion to alcohol. The resulting potential for conflicts between food and fuel has been frequently criticized. This argument has great emotional appeal—it also largely misses the point. Most American grain feeds people only indirectly, through livestock. If our concern is to increase the amount of grain available to hungry people, the solution is not to criticize a grain-based alcohol program but to stop feeding grains to our pigs, chickens, and cattle.

Even better, however, is to run the biomass program on wastes. Many attractive feedstocks are currently a disposal problem. In California, for instance, rice straw is now burned in open fires. Used as a biomass feedstock, it would, coincidentally, solve a major air pollution problem. Each region has its example of a potential biomass feedstock, from apple pumice in Washington state to energy studies in Washington, D.C.

The diversity of waste feedstocks makes it hard to estimate their total biomass potential. The cotton-gin trash currently burned or dumped into wetlands in Texas would be enough to run every vehicle in Texas at present efficiencies. There is enough diminished-quality grain in an average Nebraska harvest to run one out of ten cars in the state at 60 mpg for a year. Walnut shells in California, peach pits in Georgia, food-processing wastes in most places—each by itself is nationally insignificant, but it might be of great importance locally. Logging wastes can be used to make thermochemical methanol and, with emerging technology, perhaps ethanol directly; with other large sources of biomass, many local wastes can yield the required five to six quads per year.

Urban wastes represent a vastly underused resource. Municipal solid waste has both attractions and technical problems. Generated at the rate of 136 million tons per year in 1980 (estimated to be 250 million tons per year by 1995), this waste could supply several quads per year if combusted. The Congressional Office of Technology Assessment estimates the heat yield of currently combustible municipal solid waste at two quads per year. Recycling the inorganic waste could save an additional one quad of energy per year: the solid waste of cities contains two-thirds of the U.S. consumption of glass and paper per year, one-fifth of the aluminum consumption, and one-eighth of the iron and steel consumption. Careful conversion to methanol or other liquid fuels, rather than mere combustion to make heat, may be even more attractive and may avoid much of the risk of inadvertently creating toxic wastes (such as dioxins) when burning the plastics and treated papers with which this solid waste abounds.

An interesting component of municipal solid waste is urban forestry waste, such as tree trimmings, grass cuttings, and cleared brush. The Forest Service estimates that 150 million tons of urban tree wastes are generated each year. In Los Angeles County alone, 4,000–8,000 tons of separated woody material go into the landfills each day—a waste of about 1,000 megawatts thermal. Some cities deliberately manage their urban forests for wood production. Like municipal solid waste and sewage, urban forestry waste amounts to a substantial annoyance waiting to be converted to a resource. Among the technical options for converting such wood wastes are new gasifiers producing very high yields of methanol (83 percent in one recent design) or even making gasoline directly, as Dr. Tom Reed of the Solar Energy Research Institute told us in 1982.

There are several possible exceptions to the proscription against special crops. Among the most attractive potential biomass crops are plants of the family Euphorbiaceae, including spurges, cassavas, and poinsettias, for example. They come in hundreds of varieties, adapted to conditions ranging from deserts to rain forests. They share a sap rich in various resins, terpenes, and other fuels or fuel feedstocks. One study concluded that *Euphorbia* planted on a land area equal to that of Arizona could meet one-fourth of current U.S. petroleum consumption. *Euphorbia* grows well on otherwise marginal lands and has minimal water requirements. With careful attention to the impact of such species on the soil structure of sensitive lands, this could provide an attractive second crop in many regions.[31] Other dryland crops might offer similar potential. Another possibility for special cropping is cattails. Douglas Pratt at the University of Minnesota has shown that sustainable, ecologically sensitive cropping of cattails could yield just under one quad of liquid fuels per day—up to one-fifth of ultimate national needs for vehicular liquid fuels.

A VEHICLE FOR REFORM

It is becoming increasingly clear that current agricultural practices are unsustainable and that failure to reform those practices could leave us short not just of fuels but also of food. Remedies are less clear: we know more about what does not work than about what does. However, a properly designed biofuels program might serve to reduce, not increase, pressure on agriculture. The residues from bioconversion should be put back on the land. Even thermochemical conversion can preserve many trace elements and wet-chemical or bacterial conversion can save everything in the feedstock except the hydrogen and much of the carbon.

Many farmers were eyeing the Gasohol program as a possible way out of their financial straits. But without basic reforms to overcome the myriad problems of "modern" agriculture, such farmers have been keenly disappointed. We need to know a great deal more about the problems inherent in agriculture, and what effect a biomass fuels program could have on land and the economics of agriculture, before we jump into a comprehensive biofuels campaign. We need to know which feedstocks are most appropriate and which conversion processes are the most cost effective and on what scale. Most important, we need to know whether agriculture can be reformed so that it could sustain a biofuels program. Until the sustainability of a biomass fuels program is established, highly efficient vehicles should remain our first priority to save liquid fuels. An efficient fleet of vehicles on the roads would provide us with as much energy-supplying capacity as any alcohol or synthetic fuels program, and it would do so more quickly, more cheaply, and with less damage to the land.

The issues of economics and ecological destruction are forcing American farmers to reevaluate their methods of providing this country with food. Energy scarcity has become the most visible reason for change, yet in the long run, other factors such as soil destruction and toxic-substance accumulation will prove more persuasive. As we turn away from the failing, conventional methods of agriculture, we should be searching for solutions that are secure and environmentally sound. Our fuel, our food, and our future could depend on it.

7

Industrial Versus Biological Traction on the Farm

Marty Bender

It might seem that American farmers, at least in the early stages of a declining energy future, wouldn't have to worry about ample supplies of energy. Farms consume only 3 percent of all U.S. energy today, and since food is top priority, one would think that government policy would guarantee every farmer a necessary allotment. This assumes that such virtues as wisdom, common sense, goodness, and efficiency will be exercised by bureaucrats to treat all farmers equitably. A more pessimistic point of view, however, might be more realistic. There are many uncertainties in our energy economy now: the future will probably be even less predictable.

The problem the future farmer will have is ages old: passing on the costs of production. As the price of energy increases, the farmer will be even more pressed than he has been; if he is a small producer, he may be

This chapter grew out of a paper, "Horses or Horsepower?" that Wes Jackson and I co-authored in 1982. Without his contribution, this chapter could not have gotten as far as it has. I am very grateful for the questions and proposed answers by Charles A. Washburn, professor, California State University, Sacramento, and by Wendell Berry. Further thanks are extended to Maurice Telleen, publisher of *The Draft Horse Journal*, Waverly, Iowa.

squeezed right off the farm before the palsied hand of bureaucracy can act. It seems prudent to meet future agricultural energy requirements by growing all farm traction energy. Currently, the best example of a renewable fuel technology for the mechanized (tractor-powered) farm is the ethanol still. And the best example of a horse-powered farm is that of the Amish.

We can grow crops that can be converted into ethanol to be burned in today's tractors (with a few modifications). At present, it appears that the economically viable still will be not an on-farm still but a community still capable of producing one to two million gallons of ethanol annually. To make a profit, such a still must not only have all of its leftover wet stillage fed to an adjacent feedlot of at least several thousand head of cattle but must also receive government subsidies.[1]

Or, we can grow crops that can be fed directly to draft animals as we did on a broad scale less than fifty years ago. At present, in the Amish culture, horse-powered agriculture is still thriving. The Amish have doubled their population in the past generation and most have stayed in farming.[2] The general farm population of the United States is now one-fourth that of 1950, and the number of farms is now four-tenths that of 1950. In contrast to ethanol stills that require government subsidies, the horse-powered agriculture of the Amish has done well without government subsidies. In the face of unpredictable government policy, it might be worthwhile to consider an increase in the breeding of draft horses and mules. A large number of light draft animals could be readily obtained by breeding heavy draft stallions with a large number of light mares, such as standardbreds and quarter horses. It will take a longer time to obtain heavy draft animals because there are not as many heavy draft mares as there are light mares. But a large percentage of the horses in the early twentieth century were light draft animals that could be worked in the fields or used for transportation.

When we begin to produce this traction energy, two limits will quickly emphasize the importance of overall energy efficiency. The most obvious, perhaps, is that traction equipment will be in competition with people and livestock for food. After comparing energy use by the draft animals and by the tractor, I will compare the cropland acreage needed to feed people and horses and mules in the United States. An equally serious limit is how little energy an acre produces, given the context of our present fossil-fuel economy. Once these earthly limits become conventional wisdom, many agronomic practices that seem impractical or humorous today will suddenly become serious considerations for the farmer.

Agricultural researchers have done economic analyses of farms powered by draft animals and those by tractors early in this century when the transition from draft animals to tractors was occurring. Robert E. Ankli noted that the great range of results that are reported in studies in the corn belt indicates that the ability to organize and to farm were more important in determining profitability than the decision to buy a tractor or to continue relying on horses. No comparison in terms of energy use was done for farms or traction energy during that time.[3]

Sociologists and geographers have produced a number of economic and energy-use studies on Amish agriculture over the past two decades. Johnson, Stoltzfus, and Craumer compared Amish farms with conventional mechanized farms in terms of energy use. There were many variables besides traction energy, so farming methods, not traction energy, were being compared.[4]

Some farmers believe that draft animals are a more appropriate form of traction than tractors because they acknowledge the fact that farms, like ecosystems, have biological rhythms. Some of these farmers have spent much time and effort to observe and write about how animals on a farm enable a farmer to utilize its biological aspects to his advantage.[5] Also, draft animals require a diet, which provides an incentive for a farmer to diversify and practice crop rotation. Tractors need only calories or energy and thus fit poorly into the rhythms of a healthy farm. Tractors allow farms to leave behind many biological and cultural interactions and assorted inefficiencies. In other words, the tractor forces the farmer into the industrial age now characterized by high interest and inflation. In such a capital-intensive world, farm family assets are transferred to distant places, usually cities. These assets could provide greater economic benefit at home if they could be retained within the community.

In spite of all these considerations, studies, and writings, agricultural researchers are doing far less research now on draft animals than on tractors and ethanol. And agricultural researchers have avoided studying Amish agriculture. They simply believe that cropland will not be available to feed more horses and mules beyond those present in the United States now. Also, the emphasis in agricultural research has been to reduce the number of human hours in the field by using the tractor. This is not a socioeconomic good if teenagers are denied the opportunity to learn farming skills and have to take jobs at drive-in theaters or fast-food places in nearby towns.

Agricultural researchers also believe that a tractor uses less energy than an appropriate number of draft animals simply because a tractor can be shut off between jobs, whereas draft animals eat while they are idle. However, they have not considered the energy required to manufacture and maintain the tractor or to replace the draft animals. Moreover, the energy required to build the still and to produce the ethanol must be calculated.

And energy credits should be given for the draft animals' manure and for the wet stillage left over from ethanol production. Another consideration is that the draft animals will pull all the harvest equipment, such as the corn picker and combine, but as in Amish agriculture, the threshing, cutting, and picking will require a small engine. Such an engine uses an estimated 75 percent of the total energy used by conventional machinery in the harvest; draft animals can provide the remaining 25 percent of traction energy needed in the field.[6]

A key consideration in making a comparison is the number of acres that can be farmed by a number of draft animals compared with the acreage that can be managed by a tractor of a given horsepower. Nowadays, a typical farm has from 320 to 640 acres of cropland that can be tilled with a 125-horsepower tractor, and such a farm also has a second tractor of about 70 horsepower for backup. So the mechanized farm will be represented here by two examples, a 320-acre farm and a 640-acre farm, each using two tractors.[7]

The horse-powered farm will be represented by three examples. The first is the Amish as they farm currently. Hostetler states that the average size of Amish farms in Lancaster County, Pennsylvania, was eighty-four acres in 1978 and that a farmer normally has six draft horses. An Amish farm usually has about twenty acres of woods and permanent pasture, which leaves about sixty acres of tilled land for an average of about ten acres per horse.[8]

The second example: in the United States in 1920, the horse and mule population was about twenty-two million work stock and three million colts and fillies. There were then 360 million acres of harvested cropland, which suggests an average of 16.4 acres per work stock. However, unknown millions of those work stock were not on the farms but in cities like New York and Chicago. The era of the automobile was young, so horse transportation in cities was still common. This suggests that the average should be greater than 16.4 acres per work stock on the farm. Moreover, there were about 400,000 tractors during the early 1920s. At that time, the tractor could perform only certain operations, since the general-purpose tractor had not come into use yet. An Illinois Agricultural Bulletin in 1921 listed planting, cultivating, and husking corn; hauling manure; mowing, threshing, hauling, raking, tedding, and putting hay into the barn or stacks; and picking up corn after the binder as examples of operations that the tractor could not yet perform. Considering this, I will be conservative and round the 16.4 plus acres down to 15 acres per work stock. Thus, six horses could work ninety acres for this example.[9]

Horse farming in Iowa during 1930–1931 is the third example. Ankli states that there were 4.2 work horses per farm (1930–1931) on average crop acreage of 70–79 acres, 7.0 (1930) and 6.5 (1931) per farm on 120–159 acres, and 12.7 (1930) and 13.0 (1931) per farm on those over 320 acres. There were no tractors on the farms in this survey. This suggests an average of about twenty acres per horse. For this example, six horses could work 120 acres.[10]

A conventional comparison of horse-powered and mechanized farms would have the former growing a crop mix to provide a diverse diet for the horse and the latter growing only corn because the tractor needs only ethanol. Because corn produces many more calories of energy per acre than other crops do, the mechanized farm would have a higher ratio of energy input and output than the horse-powered farm and thus would be considered the more desirable method of traction. However, any piece of land that has the same high-yield crop, year in, year out, will require more fertilizer and more pesticides: more fertilizer because runoff will be worse than if crops are rotated and more pesticides because pests build immunities. In the long run, growing corn on the same piece of land, year after year, will prove to be unsustainable because of the soil erosion (or the heavy use of pesticides in no-till corn).

Because I am interested only in farming that is sustainable, I will have both farms grow the same crop mix in my example—a crop mix typical of Iowa farms in 1940. A quarter-section had 150 tilled acres growing 60 acres of corn, 30 acres of oats, 30 acres of soybeans, and 30 acres of alfalfa. With this same distribution, the mechanized farm of 320 acres would have 128 acres of corn and 64 acres each of the remaining three crops. Such a crop mix can be grown in rotation and in strips along the contour of the fields to reduce soil erosion and pest buildup. A crop mix also reduces the impact of depressed prices or yield in a particular crop.

For this comparison, the mechanized and horse-powered farms will grow the same crop mix with the same treatments so that they will have the same crop yields. So all that needs to be calculated is the energy input per acre to provide traction. Equivalent crop yields are assumed because crops on horse-powered farms and mechanized farms receive different treatments and thus cannot be compared for yields solely in terms of traction effects. John Hostetler reports that the Illinois Amish farms yield from 70 to 130 bushels of corn per acre while the neighboring farms yield 150–170. Johnson, Stoltzfus, and Craumer, in their Illinois study, found the yields to be 115 bushels per acre on the Amish farm versus 165 bushels per acre on the other Illinois farm; different amounts and kinds of fertilizer were applied. Amish farm fertility has traditionally counted on legume-fixed nitrogen and manure, whereas other American farmers rely on a huge petroleum expenditure.[11]

Along with an explanation of the calculations for the Amish example of a horse-powered farm and for the 320-acre example of a mechanized farm, tables will show the per-acre energy input and percent of farm acreage for fuel or feed for all the examples of both farms. Then I will consider the possible conflict between animals and humans for food if farms had more horses and mules.

ENERGY FOR THE TRACTOR

For the 320-acre mechanized farm, Table 1 shows energy for seedbed prep-aration, planting, cultivation, and harvest of the four principal crops.[12]

A community still of the size discussed earlier produces 85 percent as much ethanol as a central plant can from a given weight of corn. So a bushel of corn grain weighing fifty-six pounds (7,000 BTUs per pound) is converted into 2.2 gallons of 200-proof ethanol (85,000 BTUs per gallon) and the equivalent of 16.3 pounds of dried distiller's grain with solubles (5,800 BTUs per pound) from the leftover wet stillage. Thus, 48 percent of the corn's energy ends up in the ethanol, 25 percent in the wet stillage (the given energy credit), and 27 percent is lost as heat. So 733 million BTUs of corn grain are needed to produce the 352 million BTUs of ethanol for trac-tor fuel.[13] The national average for corn grain yield is about 100 bushels per acre, so 18.7 acres of corn are needed for tractor fuel.

Energy spent in the field is not the total expenditure of energy. I must also determine the energy spent for tractor replacement. I assume here that the "principal" tractor can last fifteen years and that the "stand-by" trac-tor can last four times as long. (Of course, the same tractor won't be around the farm for sixty years. There may be no distinction between "principal" and "stand-by" tractor; if there is, the stand-by may be a tractor purchased

Crop	Acres	Million per Acre	BTUs Total
Corn	128	1.2	154
Oats	64	0.5	32
Alfalfa	64	1.5	96
Soybeans	64	1.1	70
		Total	352

TABLE 1. Tractor Fuel Consumption for Four Crops on the 320-Acre Farm

long ago but retained rather than traded in on a newer one. There are numerous possible combinations and conditions.)

Since the 125-horsepower tractor weighs 13,300 pounds and 38,000 BTUs are required to manufacture a pound of iron, 505 million BTUs are needed to manufacture the 125-horsepower tractor. This energy spread over the lifetime of the tractor, which is assumed to be fifteen years, amounts to 33.7 million BTUs per year for manufacture. Annual repair adds an additional 6 percent to this cost, or 2 million BTUs, thus giving a total of 35.7 million BTUs. To pay this annual manufacturing and repair cost with ethanol, 1.9 acres of corn are required. The 70-horsepower backup tractor weighs 9,200 pounds. Spread over its theoretical sixty-year lifetime, the annual manufacture and repair of the backup tractor requires 0.3 acres of corn. The total acreage needed for both tractors is 2.2 acres.[14]

The field work and manufacture and repair of both tractors require 20.9 acres of corn, which will yield 4,605 gallons of ethanol. To produce this ethanol, energy is required for building the community still. I have assumed that this equipment energy cost is similar to that of a central alcohol plant. So the energy required for manufacturing the community still, spread out over the gallons of ethanol it can produce in its lifetime, is 2,800 BTUs per gallon of ethanol.[15] Thus, 0.7 acres of corn are needed to pay the equipment energy cost of converting 20.9 acres of corn grain into ethanol. This brings the total up to 21.6 acres. The mash credit is 26 percent of this acreage, or 5.4 acres.

A community still of the size discussed earlier requires 30,000 BTUs to produce a gallon of ethanol.[16] So the production of ethanol from 21.6 acres of corn can be fueled by burning 14.9 tons of wood (8,000 BTUs per pound) at 60 percent efficiency. I assume an average woodlot, with some maintenance, will provide three tons per acre of harvestable wood annually; thus, five acres of woodlot are needed to provide fuel for production of the ethanol. Table 3 summarizes the acreage needed for traction energy for the 320-acre and 640-acre farms; for a summary of the energy input for traction on the two farms, see Table 5.

ENERGY FOR THE DRAFT ANIMAL

Each of the six draft animals on the Amish farm of sixty acres would work approximately 700 hours per year. This amounts to seventy ten-hour days, although the work is not distributed that neatly. Since a horse working ten hours requires a significantly richer food ration than an idle horse, it is convenient and accurate to divide the working year in this way. Most good teamsters would scale the feed allotment to match the work schedule.

	Pounds of Feed/Animal for the Equivalent of:					Acres Needed per Animal
	70 Ten-Hour Work Days		295 Idle Days		Total Pounds	
	Daily	Annually	Daily	Annually		
Corn	5	350	—	—	350	0.06
Oats	10	700	—	—	700	0.44
Hay	15	1,050	15	4,425	5,475	1.37
Total						1.87

TABLE 2. Feed Consumption of One 1,500-Pound Horse or Mule

Table 2 is an estimate for the annual feed demand for a 1,500-pound horse or mule working 700 hours.[17] For corn, I assume a yield of 100 bushels per acre, or 5,600 pounds. There are thirty-two pounds per bushel of oats, and a yield of fifty bushels per acre is not unreasonable; therefore, an acre will yield 1,600 pounds. I assume an acre of hay will yield two tons, or 4,000 pounds. The total acreage is 1.87 acres, so six draft animals need 11.2 acres of feed annually.

In addition, as mentioned earlier, small engines provide 75 percent of the harvest energy and the draft animals provide the remaining 25 percent to pull the harvest equipment around the field. The acreage of corn for ethanol to fuel the small engines in the harvest of the crop mix on the sixty acres is 1.48 acres. The capital equipment energy cost for the on-farm still to convert this requires 0.05 acres of corn, giving a total of 1.53 acres. The wet stillage credit is 25 percent of this, or 0.4 acres; 0.4 acres of woodlot are needed to provide the fuel for production of the ethanol from 1.53 acres of corn.

One of the six draft animals in my example will need to be replaced every 2.5 years on the average (horses average a fifteen-year work life). The energy cost for replacing a 1,500-pound draft animal can be determined by calculating the extra feed a mare requires from the time of breeding until the foal is born and the feed costs of the foal from birth until age two, when it can begin work.

The pregnant mare will consume no extra hay, but there is an increase in the amount of oats and corn she consumes. Her additional feed requirements during the 345 days of gestation are 1.3 acres of oats and 0.2 acres of corn. Once the foal is born, the mare will have additional feed requirements for milk production, but this is insignificant compared with the foal's consumption during the next two years of 2.4 acres of oats and 1.4 acres of hay.[18] So a draft animal replacement requires 5.3 acres of cropland, or 2.1 acres annually, spread over 2.5 years. Thus each workstock and replacement require a total of 13.3 acres annually.

The energy credit for the manure needs to be calculated. A horse excretes 35 pounds of manure daily, which contain 7.3 pounds of volatile solids with an energy value of 58,400 BTUs. The daily feed for a horse, as shown in Table 2, has an energy value of 166,000 BTUs. So the energy credit for the manure is 35 percent of the energy value of the feed, which works out to 4.7 acres for this farm. Table 4 summarizes the acreage needed for traction energy for the three examples of horse-powered farms, and a summary of the energy input for traction on the three farms is in Table 6.[19]

COMPARISON OF THE FINDINGS

An examination of Tables 3 through 6 reveals that although horse-powered farms can use more or less energy per acre for traction than mechanized farms, they need about two or three times the acreage on a percentage basis. This is because alfalfa and oats do not produce as much energy per acre as corn or trees do. In the most favorable horse-powered example in

	320-Acre Farm	(Including woods)	640-Acre Farm	(Including woods)
Field work	18.7		37.4	
Manufacture and repair tractors	2.2		2.2	
Equipment for still	0.7		1.3	
Ethanol production (fueled by wood)		(5.0)		(9.4)
Minus wet stillage credit	−5.4		−10.2	
Total	16.2	(21.2)	30.7	(40.1)
Percentage of farm for traction energy	5.1%	(6.5%)	4.8%	(6.1%)

TABLE 3. Acreage (Acres) for Traction Energy on Mechanized Farms

	60-Acre Amish Farm 10 Acres/Horse	(Including woods)	90-Acre 1920 U.S. Farm 15 Acres/Horse	(Including woods)	120-Acre 1930–1931 Iowa Farm 20 Acres/Horse	(Including woods)
Feed						
Six draft animals	11.2		11.2		11.2	
Replacement	2.1		2.1		2.1	
Minus manure credit	-4.7		-4.7		-4.7	
Subtotal	8.6		8.6		8.6	
Ethanol						
Threshing engines	1.48		2.21		2.95	
Equipment for still	0.05		0.07		0.10	
Ethanol production (fueled by wood)		(0.4)		(0.5)		(0.7)
Minus wet stillage credit	-0.40		-0.60		-0.80	
Subtotal	1.13		1.68		2.25	
Total of feed and ethanol	9.7	(10.1)	10.3	(10.8)	10.9	(11.6)
Percentage of farm for traction energy	16.2%	(16.7%)	11.4%	(11.9%)	9.1%	(9.6%)

TABLE 4. Acreage (Acres) for Traction Energy on Horse-Powered Farms

	320-Acre Farm	640-Acre Farm
Field work	733	1,467
Manufacture and repair tractors	87	87
Equipment for still	27	51
Ethanol production (fueled by wood)	238	450
Minus wet stillage credit	−212	−401
Total	873	1,654
Total Energy Input Per Acre	2.7	2.6

TABLE 5. Energy Input (Million BTUs) for Traction on Mechanized Farms

Table 6, the 120-acre Iowa farm, if the 232 million BTUs of crops (266 million BTUs minus 34 million BTUs of wood) were supplied solely by corn, then only 4.9 percent, rather than the 9.1 percent listed in Table 4, of the Iowa farm cropland would be needed for traction energy. This 4.9 percent would place the Iowa farm between that of the 320-acre and the 640-acre mechanized farm shown in Table 3.

Sustainability needs to be considered as well as the convertible energy yield of a crop. It will probably be difficult to find and breed a substitute feed for alfalfa and oats that can produce as much energy per acre as corn does without eroding soil as corn does. The soil erosion caused by corn or the heavy pesticide use with no-till corn may not justify the high energy yields of corn. If a lower-yielding crop than corn is used to produce fuel for the tractor, then the mechanized farms would not show as much of an advantage over horse-powered farms in terms of acreage needed to provide fuel. Ethanol from corn was used as the renewable fuel in this chapter because enough is known about it to provide realistic numbers for the mechanized farms. Methanol from wood may prove to be a more appropriate fuel for the tractor. Vegetable oils from soybeans, sunflowers, or perennial crops may also prove to be appropriate fuels. Research is needed to assess the trade-off between soil erosion and energy yields of various crops and cropping methods.

The close results for per-acre energy input for the mechanized and horse-powered farms (Tables 5 and 6) suggest that it would be worthwhile for the U.S. Department of Agriculture, land-grant universities, and other organizations to run a series of experiments in which mechanized and horse-powered farms would be compared side by side in terms of energy use, economics, effects on soil, quality and quantity of produce, etcetera. These

	60-Acre Amish Farm 10 Acres/Horse	90-Acre 1920 U.S. Farm 15 Acres/Horse	120-Acre 1930–1931 Iowa Farm 20 Acres/Horse
Feed			
Six draft animals	191	191	191
Replacement	28	28	28
Minus manure credit	−77	−77	−77
Subtotal	142	142	142
Ethanol			
Small engines	57	87	116
Equipment for still	2	3	4
Ethanol production (fueled by wood)	17	25	34
Minus wet stillage credit	−15	−23	−30
Subtotal	61	92	124
Total of feed and ethanol	203	234	266
Total Energy Input Per Acre	3.4	2.6	2.2

TABLE 6. Energy Input (Million BTUs) for Traction on Horse-Powered Farms

experiments would need to be replicated in each agricultural region of the United States. The experimental design should resemble the design presented in this chapter. First, the experiments would have to be conducted over a range of cropland acreages that the tractor and draft animals could handle. Second, as much as possible the farms should be in close proximity on the same soils and share such landscape features as slope and surroundings. Third, both kinds of farms should grow the same crop mix and rotation organically (without petroleum-based fertilizers and pesticides). If possible, both kinds of farms should be started on land that has been farmed organically for at least several decades. This might be possible by conducting the experiments on the outskirts of Amish communities. Otherwise, the land would have to be farmed organically for five to ten years to make the transition from conventional soils to organic soils. By then, crop yields may have stabilized enough that reliable data for organic farming can begin to be recorded. Also, if the experimental cropland has never been farmed with tractors, this would provide an opportunity to record the effects of tractor compaction on the soil and the crops.

IS THERE ENOUGH LAND?

Almost all agronomists and farmers believe that the widespread use of draft animals in the United States is not feasible because there is not enough land available to feed both humans and draft animals. Table 7 shows the U.S. agricultural situation in 1920 and at present. In 1920, 6.5 million farms had a population of 32 million people. Currently, there are 6 million people on the 2.4 million farms in the United States. In 1920, 360 million acres of harvested cropland, about the same as now, were farmed with 400,000 tractors and less than 25 million horses and mules. Exactly how many horses and mules were on farms is unknown because millions of the horses and mules were in the cities and no records were kept of the population's distribution. Currently, there are 4.4 million tractors on the farms. The present horse population is about 8 million, but almost all are used for recreation. The exact number of horses used for farming now cannot be estimated. Although the U.S. population is now more than twice that of 1920, it needs only a little more cropland acreage for food for humans. This amounts to 238 million acres today versus 210 million acres in 1920, or one acre of cropland per person now compared to two acres in 1920. Cropland productivity per acre has doubled since 1920.[20]

The 25 million horses and mules would nowadays require 45 million acres of feed, not the 90 million acres of cropland needed in 1920 because of the doubled per-acre productivity of U.S. cropland. A more roundabout calculation, based on the draft animal diets presented earlier for workstock, pregnant mare, and foal gives 49 million acres—too high a number because not all the horses and mules were draft animals in 1920.

So if the United States could quickly breed its eight million recreational horses and mules into twenty-five million working horses and mules, then thirty-five million acres of cropland would have to be diverted from export (thus leaving the U.S. per capita consumption of food unchanged) to give a total of forty-five million acres to feed the horses and mules. So it is possible to feed the current U.S. population and the 1920 U.S. horse and mule population and have eighty-two million acres of cropland devoted to export, with no currently idle marginal cropland brought into production.

TABLE 7. Past, Present, and Projected Distributions of U.S. Harvested Cropland Acreage (Million Acres)

Year	U.S. Population (Millions)	Million Horses and Mules (Workstock)	Acreage for Feed and Roughage			Acreage for Direct Human Consumption	Acreage for Export	Total Acreage of Harvested Cropland	Ounces of Meat per Capita Daily
			Horses and Mules	Dairy Cattle	Meat Production				
1920	106	25 (22)	90	⟵——— 210[a] ———⟶		↑	60	360	7.0
Present	234	8 (—)[b]	10	34	142	62	117	365	10.9[d]
2013	300	8 (7)[c]	19	58	260	106	0	443[e]	10.9
2013	300	8 (7)	19	58	95	106	87	365	4.0[f]
2013	300	8 (7)	19	58	95	139[g]	54	365	4.0
2013	300	25 (22)	60	58	95	151[h]	1	365	4.0

[a] The total, but not the breakdown, is known for the three categories in 1920.

[b] Almost no horses and mules are used for farming.

[c] There are no recreational horses and mules; they are all used for farming and transportation.

[d] Includes 0.7 ounces of fish.

[e] Seventy-eight million acres of currently idle marginal cropland need to be brought into production.

[f] A minimum amount of meat recommended by nutritionists—an adequate diet of 2,860 calories and 72 grams of protein.

[g] This would bring the four-ounce meat diet up to the current per capita consumption of calories and protein.

[h] This would bring the four-ounce meat diet up to slightly above the 1920s per capita consumption of calories and protein.

Note: All numbers are documented or calculated in the text.

However, I should not only look at the present but also the future, and I should use more conservative assumptions. The year 2013, a generation from now, is a good date for which to make some projections. At the current level of population growth, the U.S. population will increase from the present 234 million to about 300 million by 2013. It is estimated that the United States has three decades of petroleum and natural gas reserves, at the current level of production. We will then have to rely mostly on foreign imports of petroleum until the world's reserves are depleted in another generation or two, in the latter half of the twenty-first century, if the current level of production continues.[21] Imports of natural gas into the United States probably will not expand significantly because of the financial and physical risks of liquified natural gas. By 2013, the United States should try to have a sizable population of working horses and mules or a substantial program of renewable fuel production for agriculture, if we are to have a food supply that does not require importing foreign fuels.

One conservative assumption is that as the use of tractors, fertilizers, pesticides, and irrigation is reduced because of high energy prices in the years to come, the average per-acre cropland productivity will decline, though not back to the 1920 level. There can be more tractors than the 400,000 used in the early 1920s, and such technology as integrated pest management and hybrid seeds will still be used. The substitution of currently idle marginal cropland for prime cropland eliminated by urbanization may contribute to the reduction in per-acre productivity. By 2013, the average per-acre productivity will probably level off to about halfway between that of 1920 and now. With this projection, twenty-five million horses and mules would require sixty million acres of cropland in 2013. And in order to maintain the current per capita consumption of food for the 300 million Americans in 2013, the present cropland acreage for food eventually consumed by humans would have to be multiplied by four-thirds because of lower productivity and by the fraction 300/234 because of the population increase.[22]

With these assumptions, the cropland acreages currently devoted to domestic food, domestic feed, and exports will have to be redistributed to feed humans and horses and mules but with fewer exports. If Americans were to reduce their daily consumption of meat, a substantial amount of cropland now devoted to feed for livestock and poultry could be used to feed horses and mules. There should be no major economic or moral problem with fewer agricultural exports for several reasons: (1) Fewer exports do not mean less income for farmers because the market has not been eliminated but shifted to domestic consumption. (2) If the United States could develop efficient vehicles so that it would not have to import oil, the balance of trade would be more positive, thus eliminating the pressure for more agricultural exports. (3) Fewer exports mean that more food and feed will remain in the United States, so that more nutrients can be returned to cropland in the form of manure. (4) Exported grain rarely reaches the hungry who need it most, simply because they cannot afford to buy it. With expensive petroleum-based inputs whose costs must be passed on, grains cannot simply be given away on a large scale. So less exported grain does not automatically mean that more people in the world will starve.

Of course, the United States does have the responsibility not to simply feed other countries but to help other countries feed themselves. However, the United States must not repeat the narrow approach to agriculture science it used when it guided the research at the International Maize and Wheat Improvement Center and the International Rice Research Institute during the Green Revolution. The failure of the Green Revolution has shown that agricultural technology alone cannot solve the economic problems of the rural poor. Future approaches to solving such problems must acknowledge that poverty and hunger are socioeconomic and political in nature.

Before the projected distributions of cropland use can be assembled, as in Table 7, the current distribution of cropland use must be calculated. At present, of the 365 million acres of harvested cropland in the United States, 117 million acres are devoted to export, and 248 million acres are used for domestic food and feed. Of these 248 million acres, 62 million acres (one-fourth) are for food directly consumed by humans, and 186 million acres (three-fourths) are farmed for feed (feedgrains, high protein feeds, and harvested roughages).[23]

The breakdown for livestock into million metric tons of feed grains, high protein feeds, and harvested roughages is known from USDA figures, and the total acreage devoted to each of these three categories of feed has been calculated by Dan Luten, so a rough estimate of the acreage breakdown can be obtained. The acreage breakdown of the 186 million acres of domestic feed is as follows: 34 million acres (18 percent) for dairy cattle, 141 million acres (76 percent) for meat production (beef cattle, hogs, and poultry), and 11 million acres (6 percent) for other livestock (horses, mules, sheep, fish, ducks, geese, pets, etcetera); of these 11 million acres, 10 million acres are fed to the 8 million horses and mules in the United States, of which few are used for farming. The remaining million acres, fed to other livestock (most of which can be eaten), will be grouped with those of meat production to give 142 million acres, as listed in Table 7. Of the current daily consumption of 10.9 ounces of meat per capita, several percent of the meat from the 142 million acres are exported and are thus being counted a second time in the "acreage for export" column in Table 7. (Fish accounts for 0.7 ounces of the 10.9 for meat.) However, this second count is offset by imported meat. I assume that the United States will continue to import food other than meat, because many of the imports cannot be grown commercially in the United States.[24]

The third row is the projection for the year 2013 with 8 million working horses (rather than recreational horses) and with the 300 million Americans having the same per capita consumption of meat and directly consumed food as we enjoy now. The 10.2 ounces of meat will require 242 million acres of cropland. As fossil fuels become expensive, it would be prudent to obtain our fish from aquacultures rather than from fishing industries as they are now. If we assume that fish can convert grains, high protein feeds, and roughages as well as livestock and poultry can on the average, then 0.7 ounces of fish per capita would require 18 million acres of cropland by 2013, for a total of 260 million acres for meat production.[25] The total cropland acreage needed is 443 million acres, which will require that 78 million acres of currently idle marginal cropland be brought into production. For more agricultural exports, more marginal cropland would need to be brought into production. Considering the soil erosion that occurs when such land is brought into production, the best use would be to leave it in pasture or rangeland. Thus, acreage for exports is left as zero.

By 2013, if Americans were to reduce their per capita daily consumption of meat from 10.9 to 4.0 ounces (a minimum amount of meat recommended by nutritionists; about the equivalent of a hefty hamburger sandwich), then 165 million acres of cropland would be freed, which not only would eliminate the need for extra marginal cropland but also would provide 87 million acres for export, as shown in the fourth horizontal row in Table 7. The average daily diet in the United States currently contains 3,300 calories and 100 grams of protein. A reduction of meat consumption from 10.9 to 4.0 ounces would give a diet of about 2,860 calories and 72 grams of protein. On the average, this easily provides an adequate diet for Americans. In 1971, the average daily diet per capita for the world contained 2,270 calories and 65 grams of protein.[26]

The fifth row shows what would happen if we were to compensate for the reduced meat consumption by consuming more food grains. With a reduction in meat consumption, some of the acreage currently devoted to feed grains and harvested roughages would be converted to feed grain for export or to food grain for direct human consumption. An estimate is needed for how many acres of feed grain and roughage (to be converted into meat) can be replaced by an acre of food grain that provides the same amount of nutrition as the meat. The overall conversion efficiency of grain into retail meat can be estimated by weighing the conversion efficiencies for various livestock and poultry in proportion to the percent that each contributes to our meat consumption.[27] The overall conversion efficiency calculates to be about 11.3 percent on an energy input and output basis. The same result was also obtained on a protein input and output basis. An acre of equivalent food grain could feed as many humans as could the meat from 8.8 acres of feed grain; conversion of roughage would probably be less efficient. My conservative assumption is that an acre of food grain could feed as many humans as could the meat from 5.0 acres of feed grain and roughage. Thus, 33 million acres of food grain could bring the reduced meat diet up to our current daily consumption of calories and protein; this increases the acreage of cropland for direct human consumption to 139 million acres and provides 54 million acres of feed grains for export, with no extra marginal cropland in production in 2013.

The last horizontal row in Table 7 shows what would happen in 2013 if there were the same horse and mule population as in 1920 and if the per capita dietary intake were slightly greater than it was in the 1920s. Increasing the four-ounce meat diet to the 1920s calorie level (3,460 calories) with extra consumption of food grains would also increase the protein level to 110 grams, 15 grams more than the 1920s level.[28] No marginal cropland would need to be brought into production, and a million acres would be available for export. If Americans accepted the four-ounce meat diet of 2,860 calories and 72 grams of protein and did not consume the forty-five million acres of food grains to compensate for it, then forty-six million acres of cropland would be available for export. Or if there were sixteen million horses and mules and if Americans had the 1920s level of calories and protein, twenty-three million acres would be available for export in 2013.

A comparison of the sixth row and 1920 should be made. The 360 million acres of harvested cropland, about the same as that projected for 2013 in the sixth row, was farmed in 1920 with a farm population of 32 million people on 6.5 million farms. At least an equal number of people and farms would probably be needed in 2013. The 300 million Americans in 2013 would need 304 million acres of cropland for food and meat, about one acre per person (two acres were needed in 1920). The per-capita acreage is greater in 1920 than in 2013 for two reasons: (1) I projected the 2013 per-acre cropland productivity to be 1.5 times greater than that of 1920, and (2) the 1920 per capita daily consumption of meat was 7.0 ounces, whereas I have the average person in 2013 consume 4.0 ounces of meat and extra food grains to compensate.[29]

My calculations show that it is possible to have from one-third to all of the horse and mule population present in 1920 and still feed 300 million Americans in 2013 at or above the current per capita consumption of calories and protein, without bringing extra marginal cropland into production. In accordance with the number of horses and mules and the diet chosen, the United States could have as many as eighty-seven million acres of cropland for export. Most agronomists would disagree with my assumption that per-acre cropland productivity would decrease by one-fourth by 2013. They are confident that plant breeding and the use of agricultural technology will maintain current crop yields. If this is the case, then there would be 25 percent more food and feed than I calculated—the equivalent of having another 122 million acres of cropland for the last three 2013 scenarios in Table 7. This would make the prospects for having many more horses and mules in the United States much more likely.

The introduction of a large number of draft animals back into our agriculture will require a large increase in the number of small farms. This raises an important question for those of us who favor the geographical decentralization of people and at the same time complain about taking agricultural land out of production to build subdivisions. We applaud dividing a 640-acre farm with one homestead into eight, 80-acre Amish-style farms, although adding seven new homesteads could take 28 acres out of production. Most proponents of decentralization presume, however, that we would have an improved agriculture with the farming community from such an arrangement and that soil loss by erosion would be reduced because more people would be watching over the same area of land. Furthermore, those seven additional families will be eating out of gardens rather than fields. Since gardens are so intensively productive, the families would be using less land for subsistence than would people in cities. On balance, there is probably no lost production around the homestead of the smaller farms.

Amish agriculture is exemplary, not just because the Amish farm with horses but also because they use soil conservation techniques, while three-fourths of U.S. farmland is not under approved soil conservation practices.[30] But there are also other reasons why the Amish are successful in farming. Because an Amish farm is never located in isolation, community support is undoubtedly important in Amish agriculture. The whole reason why the Amish have been successful may be beyond the quantitative analysis that has been done here. If we are to develop a durable culture and agriculture, we should also seek out the qualitative aspects of Amish agriculture and culture that enable them to thrive.

8

The Sustainable Garden

Dana Jackson

According to the 1982–1983 "Gardens for All" Gallup National Gardening Survey, thirty-eight million American households—46 percent of households in the United States—gardened in 1982. The median size of home and community gardens was 600 square feet. For a typical investment of $20, a gardener produced $470 worth of fruits and vegetables, and the gross national home gardening product amounted to $18 billion.

People who garden range from those who grow tomatoes in pots on their patios, to those with a few raised beds in the backyard, to those who try to produce as many of their own vegetables and fruits as possible in a homestead quest for a more self-reliant life. They garden as a hobby, or because they desire fresh or unusual vegetables for gourmet dishes, or because they are concerned about residues of chemicals on purchased foods, or because gardening lowers food costs. But very few people in this country garden because it is *necessary* for them to supply their own food. Almost every kind of vegetable or fruit, either fresh, frozen, or canned, is available in the supermarket at all times. Only an affluent society built upon cheap oil can afford such a food system.

We know that this food system cannot last, because the current level of energy consumption cannot last. Eventually we will be unable to compensate for soil erosion by adding fertilizer. We will be unable to afford the costs of farming marginal land when we have urbanized prime agricultural land. We will be unable to afford irrigation costs for vegetable crops grown in arid climates. We will realize that our elaborate food processing, packaging, and distribution systems are not only wasteful, they are prohibitively expensive. At some point, continually rising food costs will cause more people to think seriously about growing their own food. For some, it will be absolutely necessary, as it has been during much of human history.

Few people have enough space around their homes to grow protein and carbohydrates in the form of animal meats or grains, although legumes and potatoes are feasible crops. But even in cities, many will be able to cultivate vegetables and fruits suitable to their climate and available space. The percentage of households involved in home and community gardens will surely rise.

We need a national effort to encourage the increase of home and community gardens now. Produce grown in backyards and empty lots could become an important component of a sustainable food system. Instead of a hobby, or a chore associated with hard times, gardening could become an ordinary, pleasant, and rewarding household duty, with many direct and indirect benefits to the community.

The problem is that home gardening, as practiced according to contemporary garden store directions and standard advice from horticulturalists, is not sustainable. Industrial gardening has evolved right along with industrial agriculture. The same large companies that profit by the misuse of chemicals, soil, and water in agriculture also profit by their misuse in gardening. One should always take a deep breath of fresh air before entering a garden store and try not to inhale while walking among the aisles filled with cartons, bottles, and boxes of garden pesticides and fertilizers. Many people have come to feel helpless without these products, promoted by the experts, as traditional garden practices are no longer passed on from one generation to another. If consumers are susceptible to advertising for gardening gadgets and chemical products, expenditures for gardening can be larger than the dollar value of food produced.

For gardening to be sustainable in the long run, people will need to learn methods of fertilizing plants and controlling insect pests that are not dependent upon the industrial model. In general, these methods add up to what we call "organic" gardening, but specific methods and practices must be designed to fit the soils and climates of particular places.

Fertilizer means one thing in an industrial garden, something else in an organic garden. Fertilizer companies have used the expression "plant food" to describe the nutrients required for plant growth. The expression is inaccurate, because plants make their own food through the process of photosynthesis. Plants with chlorophyl combine carbon dioxide with water in the presence of sunlight and produce sugars, which give them energy to grow. What plants need are inorganic chemicals such as calcium, phosphorus, nitrogen, and potassium to construct stems, roots, and leaves. In the industrial garden, special combinations of these minerals are applied as "plant food" for tomatoes, strawberries, or salad greens. The organic gardener does not fertilize particular plants, but works on building up the soil so that it contains enough minerals for plants to take what they need. Organic materials of all kinds, which can be the remains of plants, animals, and microorganisms in various stages of decay, are applied to the soil as fertilizer. Some soils may also initially require the addition of phosphate or other minerals in rock form.

In our garden at The Land Institute, we spread cow or horse manure and straw all over the soil in the fall or winter. The garden is heavily mulched each summer with hay or straw, and the mulch is tilled into the soil at the end of the season along with the remains of all the vegetable plants. During the summer we build compost piles with weeds and chicken manure. Over the years we have added other kinds of organic materials, such as old ensilage, leaves, blood and bone meals, kitchen garbage, and sawdust. Nutrients for garden plants are released from these materials as they decay, and the organic acids formed during decay dissolve other minerals from rocks in the soil, making them available for the plants to use. Some people may feel uncomfortable because they cannot quantify the exact unit of each mineral placed in the soil, but experience has shown that organic materials provide overall nutrient balance favorable to plant growth.

Organic material also improves the texture of the soil, making it more crumbly. It causes water to be absorbed and held in the soil better and helps air to flow around the roots easily. These are benefits not provided by commercial garden fertilizer. Organic garden fertilizer is usually free, or cheap, and made from renewable resources. I suspect there are enough organic materials available in what society considers "waste" to provide a sustainable source of fertilizer for all home and community gardens.

The general approach in organic gardening is to build healthy soil that will nourish vigorous plants resistant to insects. But no garden totally escapes insect pests, so certain control techniques are necessary. Insect control in a sustainable garden must be based more on knowledge than upon products purchased in a garden store. Knowing the life cycles of insect pests and some of their habits, the gardener can disrupt insect life or make garden vegetables unattractive to insects. This requires careful observation as well as some reading and talking to other gardeners, but it does not require special courses in entomology. Instead of walking into a garden store and asking the clerk, "What do you have to kill the bugs in my garden?" as I heard a man do recently, the gardener would use a combination of techniques, such as crop rotation, handpicking insects and destroying eggs, setting insect traps (yellow surfaces covered with cooking oil), and interplanting with flowers, herbs, or other vegetables that seem to repel particular insects. These techniques are sustainable insofar as they do not depend upon fossil fuel or harm the soil, but they don't always work, and human patience may wear thin. In that case, there are also specific biologically safe products such as *Bacillus thurengiensis* (a bacterium that sickens and kills cabbage worms), diatomaceous earth (a powder with microscopic needles of silica, which pierce insect exoskeletons), and rotenone (a plant-derived insecticide proved harmless to warm-blooded animals). However, these aids should not become crutches to depend upon all the time.

The sustainable garden requires a holistic health approach. The drugstore solutions developed for industrial gardens are not sustainable. This pattern must be replaced with preventive medicine. First, a good environment is prepared for fruit and vegetable plants: good soil rich in humus, adequate water, enough sun and air, and drainage. Then the garden is tended. Vegetable plants seldom produce well, no matter how enthusiastically they were planted in the spring, if they are not cared for in the summer. Weeds cannot be allowed to steal water or shade the vegetables. And, of course, the kinds of crops and plant varieties planted should be appropriate for the climate.

This holistic approach should include the mental health of the gardener. One cannot expect a perfect harvest from all crops. The produce will have blemishes: spinach may have holes, or tomatoes may have cracks, and sweet corn probably will have ear worms. Rabbits may eat the new beans, squash bugs may devastate the zucchini, and wilt may get the cucumbers. The gardener should be mentally prepared for such failures. But by planting a variety of different vegetables, instead of just one or two, the gardener can be sure something will survive. The diversity will also contribute to the overall beauty and health of the garden, as well as to insect and disease resistance.

Differences among varieties of certain vegetables are also very important. The best types of lettuce or cabbage for the Pacific Northwest are not the best types for Kansas or Georgia. Gardeners should choose plants that are heat and drought resistant if they are in an area of low rainfall or that are resistant to fungal disease if their area has frequent rains and high humidity. Some of the popular, high-yielding vegetables have been bred on the assumption that the gardener will use commercial fertilizer and pesticides, and these vegetables may not be as successful in organic gardens as some older, hardier varieties. Sustainable gardening would require that multitudes of varieties of vegetables and fruits be available, allowing gardeners to choose those that specifically fit their climates, insect problems, soils, and personal preferences. This requirement will probably need an increase in the number of gardeners who save seeds and exchange them with each other. It would be helped by a reversal of the trend toward plant patenting and the integration of large chemical companies and seed companies.

The choice of fertilizer, seeds, tools, or method of pest control to be used in sustainable gardening must be based on a deep respect for soil. Soil is not simply "dirt" or "ground" or a "growing medium"; this must be thoroughly understood. Just as persons with an intelligent concern about their health are careful about the substances they take into their bodies, those with a respect for soil are careful about the substances they pour into it and how they use it.

Sustainable gardening based on a respect for soil would require dedicated efforts at soil conservation. Topsoil removed when basements are dug would be preserved for replacement in gardens that need it or for starting new community gardens. Special procedures to prevent soil erosion around construction sites, especially in housing developments where the community members would expect to grow vegetables as well as roses, would be standard practice. In many urban areas, the soil is unsafe for growing vegetables because of the high lead content, so restrictions on automobile traffic might be necessary.

For gardening practices to be sustainable, people would need to pay as much attention to water conservation as they do to soil conservation. In some areas, low average rainfall is the limiting factor in gardening. Cisterns to collect rain water, heavy mulching, planting in depressions to capture and hold rainfall, and the use of household graywater can extend supplies. In other areas, the quality of water, not the quantity of rainfall, is of most concern to gardeners. We have learned that irrigation water free of toxic chemicals is not something we can take for granted. More and more contamination is going to be discovered now that the public is alerted to the dangers, and even if we strictly enforce restrictions on the production and dispersal of toxic substances, old residues will continue to plague us in the future. It is going to be hard to develop a sustainable food system without also making the other systems providing goods and services for humans sustainable.

If there were more gardens in the United States, would we use more water? I doubt it, for many of those gardens would replace lawns. When we lived in Sacramento, California, in 1972, I was amazed at the streams of water running by the curbs. Even though it did not rain from May until October, people took the bountiful American and Sacramento rivers' snowmelt water for granted and habitually overwatered their lawns. Water wasn't even metered—each household paid a flat charge each month. There were only two vegetable gardens on the street where we lived; the backyards were occupied, instead, by dogs as large as calves and by kidney-shaped swimming pools with gas heaters. More practical use of these backyards would be to make them "edible landscapes," combinations of fruit trees, berry bushes, and vegetables, which could thrive because of abundant water and a mild climate. At the very least, the mulberry trees that have no mulberries and the strawberry plants that bear no strawberries could be replaced with fruit producers.

During World War II, the federal government urged people to plant victory gardens, and many did so. After the war, scarcities ended, and the affluence of the 1950s and 1960s made gardening seem a waste of time; with rising incomes, food seemed inexpensive. Large, "efficient" vegetable farms, refrigerated trucks, and supermarkets took the responsibility for food production away from home gardeners. But food prices rose dramatically with inflation and the energy crisis in the early 1970s, and there was a renewed interest in gardening. The rise and fall of the Consumer Price Index can almost be traced by looking at the number of gardeners. The "Gardens for All" survey showed that 49 percent of U.S. households gardened in 1975, when we were going through oil-price shock. This percentage fell to 41 percent by 1978 when the energy crisis eased and food prices rose more slowly; as we saw, this same survey had gardening up again in 1982, when unemployment was raging.

A sustainable food system for the United States must include a larger percentage of people who grow some of their own food in home or community gardens. The survey shows that in the past ten years, this percentage has varied according to food prices and the general state of the economy. But I think the percentage will slowly rise between now and the end of the century. To accommodate this need for home food production, communities should encourage beginning gardeners with training and assistance. People should be able to receive information from county horticultural extension agents suitable for the home garden, not the industrial

truck farm. Growing food to eat should be as important as growing food
to sell, and the U.S. Department of Agriculture experts, who now mostly
serve large-scale corporate agriculture, should serve all food growers.

There will be many benefits to a community when more people are engaged in growing food. Not only will families with gardens have fresher
vegetables and fruits but so will others in the community. One result of an
increase in the number of serious gardeners could be the development of
local market gardens and small truck farms, for some people would expand their gardens and have excess produce to sell.

It would be helpful to home gardeners if more local people became market gardeners. A body of knowledge from experience could be distributed
as the produce is sold. In the 1950s, people came to my in-laws' farm near
Topeka, Kansas, to pick their own strawberries for jam and freezing. But
they also came at other times to buy strawberry plants, rhubarb roots, or
other vegetables and fruits, and along with their purchases, they took home
advice on propagating plants and answers to all kinds of garden questions.

A food system that encouraged local production would not rule out specialty crops in particular areas. Kansas could not produce oranges and al-
monds but would still import them from California and Florida. However,
as transportation and processing costs rise with oil prices, the variety of
locally grown foods would increase. In a sustainable food system, people
would eat fresh foods in season; they could not have anything they wanted
at any time of the year. Home gardeners would plant those fruits and vegetables most favored by their families and preserve them for winter, so that
in February they would eat homegrown frozen strawberries instead of
fresh strawberries from Southern California.

Home gardeners could extend the growing season of some of their favorite foods by the use of solar greenhouses and solar growing frames. The
intense interest in solar energy during the past ten years has resulted in a
profusion of experiments with passive solar structures, and a wide range of
designs is now readily obtainable from books and magazines. The manufacture of high technology glazing materials so popular with architects,
whose designs are featured in *Solar Age*, probably is not sustainable. But
simple greenhouses made of glass, wood, and rock can be attached to
homes or community buildings. High energy costs for making Portland
cement will force builders to use concrete more sparingly, and the favorite
insulating materials of the 1980s may be replaced with cheaper, and, I suspect, less effective substances. But basic materials will be available, some
recycled, to construct small greenhouses.

Although the dream of large, ripening tomatoes and cucumbers in February, while the snow flies outside the solar greenhouse, cannot be fulfilled,

some vegetables could be harvested in almost every month. Spring lettuce, spinach, and radishes could be ready earlier, and tomatoes and peppers could last late into the fall. At The Land Institute, we picked Chinese cabbage from the solar growing frame throughout the winter of 1982–1983. When the days are just too short and cold for plants to grow in the greenhouse or growing frame, those who still want to grow food could make sprouts in their kitchen. If people felt responsibility for growing their own food, that they were producers in the national food system and not just consumers, all kinds of ingenious schemes for year-round vegetables would be devised.

Greenhouses capture our imagination. Practically everyone enjoys thinking about warm, sunny spaces profuse with green plants during the cold winter months. Small-scale solar greenhouses can provide these good feelings, as well as contribute to a sustainable food system. But people also get excited about large greenhouses, sophisticated, complex structures that shut out the harshness of nature and provide an environment totally controlled by humans. The Walt Disney people understood this when they designed the Experimental Community of Tomorrow (EPCOT) in Orlando, Florida. One of the six theme pavilions depicting the future is The Land Pavilion, built by the Dart and Craft Company, Inc. (EPCOT was vividly described in *Food Monitor's* March/April 1983 issue.) A boat takes visitors through exhibits and dioramas showing what a hard time humans have had overcoming the harshness of wilderness to grow food. Old sepia photographs show nineteenth-century food production, and black and white films reveal the backwardness of early twentieth-century agriculture. Then, scenes of modern agriculture in full color prepare visitors for the climax. The boat glides into a huge sophisticated greenhouse called the "Future Farm." In what is termed a "controlled environment agriculture," plants do not grow in soil but in sterile sand, with a drip irrigation system "feeding nutrients to their roots." Visitors are impressed that the scientists of the large food corporations can produce food in greenhouses without getting dirty. The dazzling hydroponic technology is supposed to convince the public that companies like Kraft are ready and able to take care of our food supply in the future. It is all based, of course, on the assumption that energy and materials will be so cheap and plentiful that artificial environments can be built all over the world.

But in a world where resources diminish as the number of people to use them increases exponentially, can we really expect the future to be as Kraft depicts it for us? We know those high technology greenhouses cannot produce cheap food. The United States could become like a Third World nation, in which agricultural workers produce crops only the wealthy can afford. Instead of growing basic grains to feed the nation, the growers would plant "luxury" crops in greenhouses to export to the rich all over the world.

However the future in agriculture unfolds, I think more people will find it necessary to produce some of their own food in home or community gardens. And if we are serious about making the transition to a sustainable agriculture, we should start now to increase the percentage of households that grow fruits and vegetables. Whether people are ready to take more control over their food supply is a big question. If they continue to be seduced by television advertisements to choose food by appearance, not nutrition, to choose processed foods that save them time, instead of whole foods that require creative preparation at home, our food system will continue to be dominated by industrial agriculture. And then, the soil, water, and energy will surely be exhausted, and future generations will have difficulty feeding themselves. Those who think we can all just ride the Kraft boat into the future where "controlled environment agriculture" provides our food have been sadly misled.

Once people are convinced that they should fend for themselves and grow their own produce, will they be able to do it? I think most can, if they have access to plots of suitable land, are properly informed, and especially if they are motivated by the high cost of food. My father assumed that everyone who had a large enough yard should garden. He looked down upon people if they were hard-up financially, yet were not helping themselves by gardening. Now that people have given up responsibility for growing their own fruits and vegetables so willingly, it may take years to restore gardening as a cultural value. I have no doubt that people can learn how to garden, but we must make sure that they learn methods that will make home gardening sustainable.

Gardening in ways that are sustainable would not be "going back" to the practices everyone followed before pesticides and fertilizers. We can take advantage of all that has been learned about soils, insects, climate, plant breeding, and plant physiology in order to be more successful in growing a wider variety of fruits and vegetables. Although the benefits of experience and confidence in gardening are not now passed from one generation to another as they used to be, this sort of continuity could again become a part of the culture. The potential benefits from teaching gardening to a large public by means of publications, films, classes, and organizations are enormous. When large numbers of gardeners, with a deep respect for soil, adopt organic gardening methods in their own backyards, we will be well on our way toward a sustainable food system for the whole nation.

9

The Economic Structure of a Sustainable Agriculture

Marty Strange

No doubt much of this book will provoke those who subscribe to the customary conclusion that any shift toward a sustainable agriculture means a return to subsistence agriculture, the homestead, and universal self-sufficiency in food, shelter, and clothing. Such critics might tell you that agricultural reform movements would have every farmer farm with a span of mules, a cow, a sow, and a walkin' plow. Such critics warn farmers that a sustainable agriculture means a return to drudgery, a foreclosure on community involvement, and no vacation—not even for a weekend football game at the state university.

Such criticisms are, of course, nonsense. They are a sort of death rattle on the part of vested economic interests, which fear the prospect of farmers who are economically free as well as environmentally responsible. Farmers will be just that, if they choose a sustainable agriculture.

To be sustainable, agriculture must sustain the people who are part of it. At present, agriculture ruins a substantial number of lives each year. Financial ruin is obvious, but present agriculture depletes people in other ways as well, as evidenced by the growing concern about emotional stress among farmers and farm families. The objective of an agriculture that is sustainable must be to nourish a renewable pool of human land stewards who earn a healthy living by farming well.

What such an agriculture would "look like" depends upon the cultural values of the society of which it is a part. The organizational structure of agriculture is a function of the socioeconomic structure of society, a reflection of the dynamics of decision making, power relationships, and rules of order. Agriculture, like everything else, must tick. I could describe a sustainable agriculture within a society where collective enterprise was prevalent, or where exchange was by barter, or where extended families were the rule. To take any form of society and conclude that a sustainable agriculture is inconsistent with it, is to conclude that such a society is self-destructive, even suicidal. This conclusion is precisely what some of modern American agriculture's most severe critics claim.

I prefer to believe that commercial agriculture can survive within pluralistic American society, as we know it—if the farm is rebuilt upon some of the values with which it is popularly associated: conservation, independence, self-reliance, family, and community. To sustain itself, commercial agriculture will have to reorganize its social and economic structure as well as its technological base and production methods in a way that reinforces these values. Right now, agriculture is undergoing a structural transformation in the opposite direction. It is being remade in the image of industrial society. Agriculture is completing its industrialization just as that era is coming to an end.

CHARACTERISTICS OF INDUSTRIAL AGRICULTURE

Industrial production relies upon standard procedures and precision technology to achieve uniform results in mass quantities. The biological propensity for diversity is an obstacle to industrial procedures. Discretion on the part of workers is, therefore, to be avoided. Industrial enterprises are organized to minimize worker discretion and so, increasingly, is agriculture.

The principal organizational characteristic of industrial enterprise is the separation of ownership from operation. There are owners (investors) and workers, plus managers who run the affairs of both owners and workers. The corporate form of business organization insulates owners and managers from responsibility for their actions and limits their financial liability. Corporations are almost universal. The principal means of finance is borrowing, and debt is regarded as perpetual. The industrial purposes are unambiguous: profit, growth, accumulation of assets. Aesthetic and social values are so alien that they require separate recognition under the rubric of "corporate social responsibility."

American agriculture has increasingly emphasized industrial values and assumed the characteristics of industrial organization. Farms have become more specialized, relying on the prescriptive application of standard technologies, producing on a large-volume basis, using sophisticated machinery that limits the farmer's ability to adapt or change. Farms increasingly rely on perpetual debt to foster expansion. Most important, as competition for land among expanding farms increases land values, the tendency is to separate farm ownership from farm operation. This trend toward industrial or factory organization is evident on nearly every farm in America.

For example, it used to be a satisfactory lifetime accomplishment for a farm couple to pay for their farm, provide for their own retirement, finance the entry of one child into farming, and send the other children through college. Now farmland has become so valuable that merely to send the children through college is virtually to disinherit them. A college education is worth nothing like so much as a quarter section of Iowa farmland. A modest farm estate in Iowa can be worth well over a million dollars. All manner of devices are sought by which every heir can acquire a portion of the farm without breaking it up as a farm operation. Most frequently, a family corporation is established so that the heirs can share the inheritance as stockholders. The one heir who farms the land then becomes the tenant of a corporation composed of himself or herself and his or her siblings. The stockholding heirs move away, they marry, they change. In-laws become involved, especially when divorces and deaths alter family relationships. The family farm over several generations becomes less a personal property than an object of corporate intrigue.

This transition from agrarian to industrial agriculture is pervasive and incremental, but it is not yet conclusive. It is true, as apologists for industrial agriculture say so eagerly, that agriculture is a business and no longer just a way of life. But it is not yet true that agriculture is like any other business and no longer a way of life as well.

What is troubling about this transition is that the purpose of industrial organization is different from the purpose of agriculture. Industry exists to convert raw materials into finished products, leaving behind waste products and by-products. It is inherently consumptive and wasteful. Applying industrial techniques to agriculture requires agriculture to treat natural resources as consumable raw material. A corporate farm that has invested in irrigation in the Great Plains has a planning horizon precisely as long as it can project its return on investment, which is generally as long as the depreciable life of its irrigation equipment—or about fifteen years. It has no grandchildren who will rely on the groundwater it is mining or who will earn a living from the delicate rangeland it has put to the plow and subjected to the erosive winds. When warned that the groundwater level is dropping, the corporate planner wants to know if the water will last as long as the irrigation plumbing.

To be sustainable, agriculture must be organized economically and financially so that those who use the land will benefit from using it well and so that society will hold them accountable for their failure to do so.

CHARACTERISTICS OF SUSTAINABLE AGRICULTURE

Any social system must allow for diversity among its individual members. What makes it a "system" is the collective expression of values, the commonly held goals of the participants, and the central tendencies in their behavior. To the extent that most of the individuals can realize their goals, the system may be said to work. To the extent that only the exceptional succeed, the system fails. The glue that holds the system together is its consensus about how things ought to be, not necessarily how they are. Commonly held expectations, both in realization and in disappointment, make a social system—not the uniform behavior that provides caricatures for comedians and stereotypes for bigots.

Accordingly, farms in a sustainable system of agriculture will not be uniform in organization or rigid in structure. Particularly in a democratic society, there must be opportunity for full expression of individual choices, acceptance of responsibility (including the right to fail), and freedom to interpret the world differently from the folks across the fenceline. But some expectations remain at the core of a healthy or sustainable agriculture's economic structure. Among those expectations are the following:

1. *Farms are family centered.* The farm is both a place of work and a home. Children are raised in an environment in which useful work is expected of them and in which responsibility is not learned abstractly, but is accepted in the normal course of growing up. Learning to farm is a matter of apprenticeship; while formal education is not spurned (in fact, it is highly regarded), it is not a substitute for the practical experience of farming and the common sense derived from it and necessary to it.

Farms are family centered because the family is a logical unit of production within which to transfer skills and to provide intergenerational continuity in the farm's management. Skills, values, and success are the human "wealth" of the farmer and the inheritance of the farm child. By contrast, there is no absolute hereditary right to the material wealth of the farm, primarily the land. A hereditary agriculture in which farm assets are accumulated and passed on to heirs as a matter of right, whether they choose to farm or not, and whether they are able or not, is inconsistent with our long-held belief in economic opportunity based on meritorious performance. The ability and willingness to work the land for a living should be the principal attributes of those who farm (and of those who own farms), not the accident of birth and parentage. Of course, those who are born and raised on a farm will have a material advantage over others, but the advantage can be overcome by the others through work, and the capacity for work is the one economic factor with which we are all born.

2. *In our society, a sustainable agriculture should also be owner operated.* That system of ownership is within our experience and encourages (imperfectly) the responsible use of resources. The owner-operated system of agriculture—and entrepreneurship of all sorts—gave the New World its global reputation as a place of opportunity for all. Not everyone, of course, starts out as an owner-operator, nor does everyone succeed in becoming an owner-operator. Tenancy, however, is only an appropriate vehicle for entry; it is not a permanent fate. The gravitational force of the economic system should pull land to those who work it, rather than separate it from them as do current economic policies, which reward land speculation by nonfarmers. Today, over half the farmland in America is owned by nonfarmers.

Implicit in a system that is owner operated is a stricture against owning more than one can personally farm and be responsible for. Accumulating land as an economic investment is inconsistent with an agrarian system in which farms are owner operated. This acceptance of limits on ownership comes painfully to some parts of our society, but it has always been understood as necessary in agrarian circles and among working family farmers. If there is one stereotype about farmers that farmers themselves recognize and deplore, it is that of "the guy who doesn't want to own everything, just the land next to his own."

3. *If agriculture is owner operated without being hereditary, it is because farms are internally financed.* This means simply that farm production expenses, including the cost of land, which each generation must pay, are paid out of farm earnings. Credit, of course, is an acceptable means of internal finance, as long as the debt is intended to be paid and the credit is used to finance production. The use of credit to finance the purchase of fixed assets, especially land, when the purchase is based on expected increases in their value, is speculation. Such price-inflating purchases work, as long as each succeeding buyer in a land market is willing to pay more than the last. But this is a kind of Russian roulette; when farmland is bought on credit at a price that cannot be paid out of farm income, someone eventually loses. The loser is the one who buys at the peak of a price cycle, when the price of land is least related to its current earning power. Those buyers

become farm foreclosure statistics each time an agricultural depression rolls around.

4. *Internally financed, owner-operated farms can function in a market economy only if markets are open.* In an open market, no farm, by reason of its size, can have a competitive advantage over other farms, either in the price it receives for products, the opportunity it has to sell those products, or the terms under which those products are sold. An agriculture organized to realize these expectations would be composed of farms of small and modest size, and most food would be produced on such farms. Farms would by no means be uniform in size, although large farms would be exceptions. For the most part, public policy would discourage excessive growth in a farm.

The conditions we have today in American agriculture, in which a minority of the farms (about 5 percent) produce over half the food by operating at a size that exceeds technical economies of scale, would be unaccommodated in an agriculture built on principles of sustainability. The present agriculture serves the 5 percent with policies that strangle the 95 percent. It forces every farm to grow, by granting a competitive economic advantage to expanding farms. It minimizes the risk of expansion by offering price support programs pegged to volume of production, without meaningful limits on production. It offers tax subsidies to those who invest in new machinery before old machinery is depreciated. It provides emergency credit to bail out those whose foolhardy, debt-supported expansion causes them to fail financially.

In a sustainable agriculture, growth in farm size might occur, but farms would likely grow gradually, and they would all grow more or less together. For the most part, farms would be as large as they need to be to keep the farm family fully employed using technology that improves productivity and reduces drudgery without jeopardizing resources or the environment.

PUBLIC POLICIES TOWARD PRODUCTION, TAXATION, AND CREDIT

It is always surprising how many people think that the changes that have taken place in agriculture in the past generation or two are the natural effect of some immutable law of agricultural economics. There is something in the dispassionate jargon of political economists that makes the unwanted seem necessary.

There may indeed be imperatives that will determine the look of agriculture—but they will be the product of biological, not economic, law. An economic system is only a way of organizing productive resources so that they accomplish economic purposes. It is a function of custom, culture, and social values—but it is not above the laws of ecology. We cannot have any kind of agriculture we want; our agriculture must conform to biological law. But we can have the kind of agricultural economy we want. The proper purpose of public policy is therefore to shape the economic system so that it fulfills our economic expectations without threatening our natural environment.

The policies necessary to build an agriculture based on owner-operated, family-centered, and internally financed farms are not exotic, radical, or subversive. To the contrary, they are both traditional and fully consistent with the stated intent of nearly every major piece of farm legislation adopted in the twentieth century. In many areas, new measures are needed less than the elimination of some current ones that bias agricultural policy in favor of industrial farms.

If we are to have a sustainable agriculture, production will have to be controlled within sustainable limits. The capacity of industrial systems to increase output in the short run is the principal threat to agriculture's resource base and to the competitive position of smaller farms that cannot or do not choose to use large-scale technology. Excess production destroys soil, water, and farmers. The purpose of controlling production is both to conserve natural resources and to maintain farm prices at equitable levels.

Production can be controlled only by planning, a role for which the government is uniquely qualified. The need is to match production with consumer needs. Food must be readily available at reasonable prices; farm prices must provide for internal financing of the farm operation. These needs can be planned from the top down, by experts, although many of us have an instinctive mistrust of such a process. More democratic is the Scandinavian system, in which government arbitrates negotiations between private farm interests and private consumer interests to set production goals and to determine prices. Technically shaky but socially valid, this system probably makes up in practical effectiveness what it lacks in theoretical efficiency. Thomas Jefferson's remark that self-government is always better than good government sums it up well.

Distributing the right to produce the amount needed is troublesome, but it is necessary if an equitable distribution is expected. Quota systems administered by governments are considered antithetical to free enterprise, but they have been used successfully in U.S. agriculture many times over, particularly in commodities that are easily overproduced (tobacco, for instance), or need to be protected from foreign competition in the national interest (sugar).

One of the key issues in a quota system is whether the right to produce can be transferred from one farmer to another. When they can be freely sold, allotments tend to accumulate with those who can pay for them, and the future benefit of the quota is capitalized into its purchase price. Such a quota plays much the same role in restricting entry to agriculture as do high land prices. Government can intervene by helping beginning farmers finance the purchase of the quota, but such measures really only increase the competition for and the price of the allotment. A better method is to make the allotment nontransferable or to restrict competition by limiting the list of eligible buyers. In Canada, local marketing boards distribute the quotas among potential producers; in Scandinavia, even land sales are restricted by local boards, which determine eligible buyers, selecting primarily those who need land. If limiting production is in the national interest, then distributing the right to produce among the willing public is a difficult job that someone has to do.

The most misdirected of present public policies is undoubtedly tax policy. The thrust of federal tax policy is to subsidize the substitution of capital for labor. Among an interwoven series of provisions, the most important are these:

• When capital assets—like land or breeding stock—are sold, the income is not taxed as regular income but at a much lower rate, as capital gain.

• When depreciable goods—like tractors, buildings, and implements—are bought, they can be written off taxable income at a faster rate than the rate at which they actually decline in value; this encourages more frequent and larger purchases of such goods.

• Some favored investments (mostly in specialized equipment) are given an investment credit—that is, a percentage of the purchase price is immediately subtracted from the amount the purchaser owes in taxes.

Eliminating the use of these tax deductions would do little or nothing to change the real economics of food production. It would do a lot to alter the pressures to expand that now distort the decision-making process of most American farms.

To benefit from these tax subsidies, you have to spend money, and you must have the money to spend. A 10 percent investment credit on a $60,000 tractor means nothing to somebody who doesn't have even $10,000 for a down payment. Moreover, the more money you have (that is, the higher your income tax bracket), the more a deduction from taxable income is worth to you. A deduction of $1,000 is worth $500 to a person in the 50 percent tax bracket; someone in the 25 percent tax bracket only gets a benefit of $250 from the same deduction. A progressive tax schedule means a regressive distribution of benefits from deductions. The richer you are, the more you get.

These tax breaks also encourage overproduction, a problem of growing importance in hog production, for example. This means lower prices for hog farmers who, unlike some high-income investors responsible for the tax-subsidized expansion, depend on farm income for a living. Of course, it also means lower pork prices at the supermarket, a point that apologists for such policies rarely overlook. The treachery of such logic too frequently escapes detection, for the inescapable fact is that the lower price for pork in the supermarket is offset by the additional tax burden borne by all of us to cover the tax revenue the government forgoes in its generosity to high-income investors in hog factories.

The most controversial federal tax is the estate tax. The principal purpose of this tax is to prevent accumulations of wealth from generation to generation. By doing so, it would preserve economic opportunity for each new generation. Historically, the estate tax has not meant much to farmers because their estates have not been worth much. In recent years, however, as land values have increased and as farm sizes have grown, farmers who have lived poor most of their lives (much of the time making land payments) die rich. Generally, they yearn to pass on to their children whatever they have built. Congress has been quick to grant tax credits or exemptions from the estate tax so that a farmer could pass on a family-sized farm tax free. Much controversy has surrounded the appropriate size of such a farm.

The irony is that any exemption from the estate tax tends to thwart the purpose of preventing accumulation of wealth. A farm heir (who may or may not farm) who can inherit free and clear a farm that is already large enough to be efficient can use the equity in that farm to finance further expansion. That heir has a competitive advantage over other beginning farmers who are trying to start without inherited wealth. Such inequity drives the wedge a little deeper between the ability to farm and the ability to own farmland, for the easier it becomes to inherit farmland, the harder it becomes to acquire it in any other way. Such policies may benefit those who are the heirs of established farmers (and benefit them in proportion to the value of their parents' estates), but they diminish the opportunities of would-be family farmers. In a strange perversion of the term, the "family farm" becomes a front for a hereditary, landed gentry, precisely the system of land tenure that the estate tax was supposed to prevent, and from which our ancestors (our European ancestors, anyway) fled when they came to this continent.

In an internally financed, owner-operated agriculture in which production is kept in line with sustainable limits, the right to produce is distributed equitably, and the price of farm commodities reflects their current costs, the heirs of farmers should inherit free and clear only the most valuable agricultural asset—the skill to farm. If they choose not to use those skills, they forfeit the most valuable portion of their farm inheritance. A tax on the value of the rest of their inheritance—the farmland—is consistent with a policy of preventing accumulated wealth. It would keep land on the market, and land prices would be within the means of those who farm it for a living. And there would be little incentive to own farmland if you did not intend to farm it.

If tax policies are inequitable, credit programs have proven to be only slightly more deserving of support. Historically, farm credit programs sponsored or administered by the government (or by government-sponsored cooperatives) have been justified as a means of providing entry-level capital to beginning farmers, as a means of stabilizing farm prices (through nonrecourse loans on crops), or as a means of financing the normal credit needs of family farmers. Unfortunately, as farms have become larger and more capital intensive, their credit needs have changed. As pointed out earlier, credit increasingly serves to finance almost continuous growth in the size of farms by using existing equity to secure loans to finance expansion. Leveraging existing land equity enables these expanding farms to pay more for additional land than it would be worth to them, if they had to pay for it from the income earned from farming it. Emergency credit has also been needed during periods when farm prices dip so low that these expanding farms cannot make the mortgage payment.

Government farm credit programs have been subtly but indisputably a part of this transformation in the use of credit. The Farmers Home Administration, the principal agency of federal government farm credit, was founded in the 1930s to help tenant farmers buy land and become established as owner-operators. Over time, however, its mission broadened, first to make expansion loans to established farms too small to produce a livable income for the family, then to cover financial crises attributable to natural disasters, and, finally, to cover economic emergencies caused by low farm prices. At first the agency provided loans in modest amounts only to family-sized farms that could not get credit elsewhere. Now it makes loans to farmers larger than family size who do not depend on farm income and who can get credit elsewhere but prefer to get it from the government because the terms are more acceptable. And the loans have not all been modest; some borrowers exceed a million dollars in indebtedness.

From an agency that was once a "lender of last resort" to the have-nots in American agriculture, the Farmers Home Administration has become a lender of little resistance to those who have grown rich at the smaller farmer's expense. Its function now is less to provide economic opportunity for tenants than to absorb the risk of expansion by established farms. Credit programs designed to help farmers to buy farms might be necessary to a sustainable agriculture, but a major overhaul of the present federally sponsored farm credit program will be needed.

The policies advocated here will not guarantee that anyone can prosper on 160 acres. Instead, their purpose is to strip the farm economy of the biases that now favor industrial farms while discriminating against smaller farms that practice a more sensible, sustainable agriculture. Proponents of these changes see small farms as part of the solution to the problems of an unstable industrial agriculture that destroys natural resources and makes farmers financially vulnerable to a host of parasitic industries and economic plagues. The objective here is to restore competition by removing the policies that currently provide advantages to expanding farms. The objective is not to guarantee success on 160 acres but to eliminate policies that encourage irrational growth on nearly every commercial farm. Let us see if big farms prevail in an economy not biased in their favor. Let us see who survives in an agriculture designed for sustainability.

10 Sunshine Agriculture and Land Trusts

Jennie Gerard
Sharon Johnson

What Pagosa Springs, Colorado, lacks in size it makes up for in beauty. The area is characterized by high mountain valleys nestled among snow-capped Rocky Mountain peaks. Cattle-dotted pasturelands reflect that this is a ranching community, although the scenic quality and abundant wildlife attract tourists, hunting and fishing enthusiasts, and refugees from places where cars rather than cows are the dominant feature on the landscape.

To a small town like Pagosa Springs, its desirability is its biggest threat. Demand for property there has pushed up land values. Ranchers are forced to pay estate taxes disproportionate to their earnings, and real-estate developments are competing for limited water supplies. The county has minimal zoning and regulation and, in spite of a recognized fear of losing the agricultural economy and scenic character of their home, residents are skeptical of governmental solutions to their dilemma.

Betty Feazel, a local rancher, is one of those concerned residents. She owns and operates the "At Last" cattle ranch at the base of Wolf Creek Pass on U.S. Highway 160 just east of Pagosa Springs. Her family has owned the ranch since 1922. "One of the things that worries me about growth is the distortion of the economic base when productive farm- and ranchland are idle," she says. "Colorado's agriculture contributes well over $3 billion a year to our state economy—twice the revenue brought in by mining and three times that generated by tourism. Colorado is the second largest agricultural producer among the eleven western states, topped only by California.

"Take my small ranch. On a summer pasture program, it produces about 70,000 pounds of beef a year. That is a useful contribution to the economy and to the food supply. I feel real pain when I see working ranches, particularly those with irrigated hay or crop fields, bought by land speculators who take it out of production and turn off the water while they devise plans for trailer courts and so on.

"The real value of land is not the biggest dollar you can get; it is what that land can produce for human food, fiber, and shelter to maintain the earth as a habitat to support mankind."

Betty Feazel and several ranchers of Pagosa Springs who share her values are fighting to protect their community's ranchland. Their weapon against spreading development is the local agricultural land trust.

THE LOCAL AGRICULTURAL LAND TRUST

Land trusts emerge from a long evolution of conservation techniques and a history of experiments with various methods of holding land "in trust" for a public that appreciates open space, wildlife, and scenic values. The land trust concept dates back to the establishment of the trustees of reservations in Massachusetts in the 1890s. The more immediate predecessors of the local agricultural land trust are the numerous land trusts established during the upsurge of conservation activity during the 1960s and 1970s. To date, they number about 500 nationally and protect 680,000 acres of land.

Establishers of trusts have focused primarily on preserving open space and nature sanctuaries. More recently, though, they have also become concerned about protecting agricultural land. The hybrid local agricultural land trust is distinguished from its predecessors by its local origin and exclusive focus on local agricultural needs.

Local agricultural land trusts are formed to remove development pressure from prime agricultural land. To do this, land trusts acquire interests in the land—usually through a conservation easement. This easement is a specialized conservation tool that can protect the open character of privately owned land.

Ownership rights can be viewed like a bundle of sticks: one stick represents the right to build on a piece of property, another to mine minerals, others to cut timber, drill for water, and subdivide—as governed by local land-use regulations. These rights can be conveyed individually or as a bundle. The conveyance of a full bundle of rights amounts to a conveyance of the property in fee simple. Conveyance of a selected right is called an *easement*. A common example is the right to cross over a piece of property—called a right-of-way easement. This is a positive easement; it allows the easement holder the right to exercise the acquired uses. By contrast, a conservation easement is considered negative because the easement holder is bound not to exercise rights over the land.

A conservation easement is a deed conveyed from a landowner to a qualified conservation organization (like a land trust) or a public agency in which the landowner places permanent restrictions on the use and development of his property. For example, the landowner may deed the right to construct buildings on his property to a land trust. The land then is bound by the terms of the easement. The land trust can never exercise those rights it holds to develop the land; it simply serves as the guardian of the easement. In effect, the potential for developing that land is extinguished.

A farmer or rancher who wants to ensure that his property will not be subdivided can convey his property's development and subdivision rights in the form of an easement to a land trust. Exactly what rights are conveyed and what rights remain with the landowner are specified in the deed of conservation easement. The terms typically allow the landowner to continue the historical uses of his land, while permitting construction or replacement of limited additional buildings like barns, outbuildings, fences, or corrals. Easements can range from "forever wild," which would leave the land entirely in its natural state, to much less restrictive plans that continue agricultural uses and allow limited numbers of houses and residential lots. When the landowner sells the property or bequeaths it to heirs, only the retained rights are conveyed; a new owner is bound by the same restrictions. The conservation easement over the land exists in perpetuity.

A land trust makes a long-term commitment to the lands it acquires easements over. It assumes responsibility for seeing that the terms of the easement are properly followed. When they are not, the land trust must seek legal action against the violators. Thus, the strength of an easement and a land trust's ability to enforce it depend wholly on the language of the easement restrictions and the descriptions of the present land condition collected at the time the easement is made. These "baseline data" will be used to determine if the land condition is being maintained. Consequently, easement language needs to be specific and the baseline data appropriate and detailed enough to allow for meaningful long-term comparisons.

Although the conservation easement cannot guarantee that farmers are going to want to keep farming, it guarantees at least that they will be able to do so if they wish, for the land will not be paved over or dug up or subdivided and built upon. And if farmland restricted by a conservation easement passes into the hands of an owner who does not want to farm it, the land is available for any uses compatible with the terms of the easement. In any case, that land will never become a shopping mall.

Protecting the land from development is not all that a conservation easement offers a farmer or rancher. An easement can also help with the finances—specifically, with taxes. The donation of a conservation easement to a qualified organization is viewed as a charitable gift, and the value of the gift can be used to offset the donor's income-tax liability. For tax purposes, the value of the easement is determined by first appraising the property at its fair market value without the easement and then with the easement. The difference between the two values is the financial value of the easement that is used to calculate the tax benefits. Few owner-operators need this income-tax deduction, however, because the interest they typically pay on substantial loans and the business write-offs they claim offset any income tax they owe.

More important is how the gift of a conservation easement lowers the often crippling estate taxes on a farm or ranch. Land is appraised for estate-tax purposes at its highest potential value—not necessarily what is there on the land, but what could be on the land. Therefore, the real-estate appraiser hired by the executor of an estate looks at a 1,000-acre ranch in the Rocky Mountains not as what it is worth as a ranch but what it would be worth as, say, fifty ranchettes and bases his appraisal on that figure. The resulting estate taxes often are so high that an heir simply cannot afford to inherit a working farm or ranch.

The Economic Recovery Tax Act of 1981 is intended to alleviate this problem by raising the limit that can be inherited tax free to $600,000 in 1987. Some legislators are fighting this change, however, and there is some doubt that the limits will be raised as scheduled. Even if they are, the high values of land, tools, machinery, and buildings often create estate values for farms and ranches that are well above the tax-free limit.

Ranches and farmland protected by a conservation easement cannot be developed into ranchettes or condominiums. An appraiser must therefore value that land at its worth for its restricted use—what the cows can pay. In this manner, the estate-tax bite is lightened by a conservation easement to an extent that often makes it possible for heirs to afford to inherit the farm or ranch and continue the operation.

The concern about prohibitive estate taxes prompted a ninety-one-year-old ranching mother to grant a conservation easement over 500 acres of scenic ranchland along the Upper San Piedra River near Pagosa Springs, Colorado. Mary Grimes wanted to find a way to pass the ranch along to her sixty-six-year-old son and avoid taxes so high that he would have to sell part of the ranch to pay them. The conservation easement was the answer. Betty Feazel's new agricultural land trust was there to receive the easement. With the help of the Trust for Public Land, a San Francisco-based land conservation organization that is a resource for local land trusts, the Upper San Juan Land Preservation, Inc., was incorporated as a land trust in 1981.

The concern that land could not be passed on to heirs because of high estate taxes also prompted dairy ranchers to organize a local agricultural land trust in Marin County, California. Here, just north over the Golden Gate Bridge from San Francisco, just east of the Golden Gate National Recreation Area, Holsteins on the hills are features of the landscape; this area is the traditional fresh milkshed for the Bay Area. Today, Marin County dairies supply one-quarter of the Bay Area's fresh milk products. The county does not have to lose many dairies before the support system within the community—the creamery, outlets for supplies, services, marketing systems, supportive zoning, ordinances, and political representation—breaks down and weakens the remaining dairies.

In Marin County, strong competition for land exists between traditional agricultural uses and subdivision for housing. Expensive homes on large lots are becoming common. This trend not only can remove significant amounts of ranchland from production but it also can bring in dogs that chase livestock, noxious weeds poisonous to cattle, and complaining neighbors who do not know what kind of fence it takes to hold a cow. New homeowners in Marin County's ranchland can view adjacent agricultural activity with its livestock, fertilizers, and dusty cultivation as a nuisance and further jeopardize the continued agricultural use of the land.

Marin's ranchers have two familiar concerns: (1) that estate taxes will force the sale of family farmland and (2) that young ranchers, unable to afford the increasingly costly land, will not be able to enter the business and maintain a vital dairy industry.

In 1978, concerned ranchers and conservationists asked the Trust for Public Land to help them develop a strategy to protect this land and their work. The result is the Marin Agricultural Land Trust (MALT), incorporated in 1980. MALT's sole purpose is to combat problems raised by rapid development and soaring land values. Its tools are the conservation easement and another innovative concept: partial development.

The influence of a conservation easement extends beyond the easement-governed parcel itself into the community. Because it shields subject parcels of land from conversion, a conservation easement assures surrounding landowners that the land can continue in agriculture. This assured continuity tends to discourage the "impermanence syndrome"—what happens when operators, seeing the farms and ranches around them turn into subdivisions, stop capitalizing their own operations, which they expect to sell soon enough. One protected farm or ranch may encourage continued agriculture throughout a whole region.

Partial development aims to accommodate two seemingly incompatible uses: in this case, agriculture and residential development. The idea is to develop a land-use plan that situates both uses on the most appropriate sites with minimal negative interaction between them. MALT is examining three existing ranch parcels that are up for sale and drawing up a plan that would cluster available housesites on the marginal lands while adjusting property lines on the grazing land to create two more manageable ranches. If the landowners agree, the sale of the two new ranches subject to conservation easements and the clustered housesites adjacent to the permanent pasture could make up for eliminating the development potential over 93 percent of the land. This way, the land trust harnesses the private home-building market to pay for agricultural land protection.

The Massachusetts Farms and Conservation Land Trust places use restrictions on land through a slightly different method. It works with the Massachusetts Agriculture Preservation Restriction Program, which is conducted by an agency of the commonwealth. The Massachusetts Farm and Conservation Land Trust negotiates for and purchases farmland, and then sells the use restrictions to the state; the land trust subsequently sells the restricted farmland to farmers at prices reflecting the land's agricultural value. The methods differ, but the results are the same as with the conservation easement: the threat of development incompatible with farming is removed by means of restrictive covenants held by an entity other than the landowner. Such other states as Connecticut, New Jersey, and Vermont have programs similar to that of Massachusetts. King County, Washington, and Suffolk County, New York, do, too.

LAND TRUSTS AND SUSTAINABLE AGRICULTURE

A sustainable agriculture requires both land and people to work it. They cannot be separated. Local agricultural land trusts address the needs of both the land and its people by preventing land conversion. As we have seen again and again, development pressure, attendant soaring land costs, and estate taxes can prove so onerous that they drive farmers off the land and encourage the planting of the final crop, houses. Removing the pressure for conversion serves to sustain the land base as well as the farming population on it.

Conservation easements over agricultural land also may promote good stewardship of that land. Because of a land trust's continued concern for the health and fertility of lands subject to its conservation easements, the easements themselves typically include clauses requiring the maintenance of soil fertility, water quality, and wildlife habitat. Easements that protect these elements simply preserve the status quo. A challenge to land trusts is to use the conservation easement as a tool to improve the condition of the land. This is not only a logical next step, particularly in the area of soil and water conservation, but in many regions of the United States, it is also a necessary one. The local agricultural land trust's monitoring responsibilities give it the opportunity to review management practice and to offer suggestions for improvement if necessary.

Techniques for soil and water conservation are already well established. In applying them to conservation easements, land trusts can use as a point of departure a resource already available to our farm communities—the Soil and Water Conservation Districts. These districts have ready access to the technical expertise of the Soil Conservation Service of the U.S. Department of Agriculture in determining conditions of erosion.

One method of incorporating the experience of the Soil Conservation Service is to request a conservation plan. This plan, normally prepared free of charge at the request of any landowner, analyzes land potential and problems, and makes nonbinding recommendations for uses that will protect the land's resources. A land trust could request a plan for each property over which an easement is sought. While this is not a substitute for the intimate knowledge of the long-term operator, the Soil Conservation Service can establish a baseline for soil and water quality in the event that the land passes into indifferent hands.

In Grafton, New Hampshire, the Soil and Water Conservation District is taking the idea one step further. The district itself holds conservation easements over land; it operates like a local agricultural land trust. This should not be too surprising: in many states these districts are empowered to hold land and, in certain respects, are organized like agricultural land trusts. The districts are legally constituted units of state government but are run by volunteers—local citizens who decide priorities and who plan and direct their own conservation programs. In many states, these districts also are empowered to exercise eminent domain for limited purposes, like flood control. The right of first refusal could be instituted in place of eminent domain, which is rarely exercised because it is politically repugnant. A district exercising the right of first refusal would have the first opportunity for purchase if a landowner decides to sell. A right of first refusal does not have to be an instituted right. It can be negotiated for and acquired, just like a conservation easement. In this way a conservation district, a land trust, or an individual can acquire a right of first refusal.

Land trusts also can serve in an educational capacity. They can organize "farm days"—open-houses by the farming community that allow the public to see and better understand local agricultural operations. They can educate the general public about the importance of laws that protect farmers and ranchers, like leash laws so important to livestock owners.

Land trusts also can lobby for more appropriate ordinances and laws and can fight the passage of "nuisance ordinances" that make unnecessary problems for agriculturalists. They can speak in favor of agricultural interests at public meetings. Where strong land-use regulations exist, land trusts can work on developing compromises between farmers and developers in coordination with local regulatory agencies.

Land trusts are not a substitute for good farmers, and they are not a substitute for good government planning. But they can help both. By monitoring the community's agricultural land, the trusts can build a constituency to protect that land. By ensuring that the land will not be paved over, land trusts can keep agricultural land in production.

Forming a Local Agricultural Land Trust

To hold legal interest in land, land trusts must incorporate. In so doing, each trust develops its own farmland protection goals and objectives and elects a board of directors from the local community. Board members include farmers and ranchers. In each step of its development, the new land trust is shaped by its community, incorporates community objectives, and is peopled by community participants.

Most land trusts incorporate as not-for-profit organizations, and then apply for tax-exempt status. As tax-exempt organizations, land trusts can accept donations of money and land and grants of conservation easements.

Such gifts to a tax-exempt organization qualify as charitable deductions for income- and estate-tax purposes under the U.S. tax code, allowing the donor to claim the maximum tax deduction permissible by law.

Incorporation requires filing the appropriate documents for not-for-profit incorporation with the state Secretary of State's office. Fees vary from state to state, but in most cases it costs less than $100 for processing the documents of incorporation.

Filing for incorporation with the state is relatively straightforward; approval of the application typically takes several weeks. Filing for tax-exempt status with the Internal Revenue Service is a separate procedure and is more complicated. Approval of the application to the Internal Revenue Service can take up to six months.

Since land trusts are voluntary organizations, capital needs are usually small. Most land trusts operate without staff and depend on their boards of directors, who donate their time. Small budgets have been raised by membership fees and events like picnics or auctions. Donations of cash or land also are solicited and are more attractive because of a land trust's not-for-profit status. Some land trusts, such as the Marin Agricultural Land Trust, have ambitious programs using revolving funds.

The Trust for Public Land offers assistance to fledgling land trusts nationwide. The trust was established in 1973 as a national not-for-profit organization to arrange for the permanent protection of both urban and rural open space. It specializes in enabling interested communities to acquire and protect important lands within their community.

11

Innocents Abroad: American Agricultural Research in Mexico

Angus Wright

Long before Americans abroad became "ugly," Mark Twain invented another phrase for them that in some ways may be far more apt. He spoke of himself and his fellow travelers as "innocents abroad," with all the knowing irony of someone who realizes that if he describes himself as innocent, he cannot be entirely innocent. I am reminded of four Methodist missionaries in Guatemala City some twenty years ago, "The Singing Ambassadors," decked out in checkered sports coats and flat tops, wildly innocent of their own absurdity among the gravely dignified Guatemalan Indians. Still, they were not such fools as to miss the fact that there was "a greater spiritual need in Acapulco," a place to which they hoped to return soon. Or, more charitably, we can recall the young American who allowed people hiding from the political police in Brazil in the early 1970s to stay in his apartment, not from any conviction of his own other than his naive idea that the dangers to himself must be exaggerated, for the police couldn't possibly be as ruthless as they were said to be.

Looking at the experience of American agricultural researchers and advisers abroad, one constantly encounters a maddening sort of moral innocence, and it is often hard to say what degree of wisdom or crass self-interest such innocence disguises. The long history of American involvement in Mexican agricultural development is an excellent case in point. American researchers and funding agencies have had a profound social and political effect on Mexico, yet they consistently disclaim any responsibility for anything but the purely technical aspects of their work. They also fail to see, or to admit, that their work has in fact been designed and implemented within the context of a larger political program, a program with the specific purpose of blocking other possible paths of economic and social development. The work undertaken in Mexico has special significance globally, for it is in Mexico that agricultural researchers first laid the basis for the vast changes that have occurred in world agriculture as a result of the Green Revolution.

In Mexico, the self-confident, production-minded approach of American agribusiness has been banging up against complex social and ecological realities for a long time. This conflict has created a sharp controversy in Mexico for many decades, one that has been made particularly acute by the recent crisis of severe foreign indebtedness in the Mexican economy and the consequent disastrous devaluations of its currency. Most Mexicans analyzing their nation's economic problems place considerable blame for present problems on the bias toward capital-intensive agricultural techniques, which have contributed to a rural unemployment and underemployment rate of around 50 percent for most of the past decade. Unemployment for the economy as a whole has been running around 20 percent in official figures, with several million Mexicans working in the United States left uncounted. In addition, Mexicans often tend to blame American influence for the persistent need to import significant quantities of basic foodstuffs, with resulting foreign exchange difficulties. While no one in or out of Mexico seriously contests the fact that American-financed agricultural research has led to impressive production gains in some crops, in some Mexican regions it is widely believed that these gains themselves have led to unfortunate economic distortions that make the solution of other problems more difficult. Some Mexicans believe that the basic political and cultural character of their country has been profoundly changed by the results of American-sponsored agricultural research.

In addition to specifically Mexican problems, ecologists worldwide worry that the work carried out under American auspices in Mexico has so severely narrowed the genetic base of grains that the long-term viability of new strains must be seriously questioned, raising the specter of global agricultural failures. Even though there is often a good deal of exaggeration and oversimplification in the way some of these charges are put forward, it is still noteworthy that they are virtually never seriously discussed by American agricultural researchers and analysts—who ordinarily dismiss such charges as mere political rhetoric or ignorance. After all, how can you argue with increased food production in a country where so many still go hungry? And how can scientists and technicians listen to malcontents when the Mexican government itself has consistently encouraged the research and provided infrastructure to put it to work? To close the matter, the seemingly innocent question is frequently asked, "To whom should we listen anyway, if not to the host government?" A brief look at the past few decades of American involvement in Mexican agriculture shows that this last question may not be as innocent as it seems.[1]

Mexico is generally considered the birthplace of the Green Revolution. In 1941, the Rockefeller Foundation established near Mexico City a research center primarily devoted to plant breeding that in 1961 took the name CIMMYT (in English, the Center for Research on the Improvement of Corn and Wheat). By the late 1950s, the center "had succeeded in producing several conspicuously successful varieties of wheat adapted to Mexican conditions (varieties which later provided the basis for the Green Revolution in India and Pakistan and other countries). They also achieved notable improvements in corn, other cereals, and some oilseed crops."[2] In Mexico, these plant-breeding achievements were complemented by heavy Mexican government investments in irrigation works, primarily in the northwestern states of Sonora and Sinaloa. Firms and individual private investors in the United States joined Mexicans and others, including a group of Greek growers, for example, in putting together the improved seed varieties and the government-financed irrigation schemes. They realized rapid growth in production, concentrated in areas that previously grew little because of the lack of reliable water delivery. The spectacular growth of agriculture in the newly irrigated valleys of the northwest brought a great deal of publicity, and when Mexico for a time changed from a wheat-importing to a wheat-exporting nation, the Rockefeller effort was widely applauded. The publicity value of the Green Revolution in Mexico can hardly be overemphasized because, from the government's point of view, it was the political value of the effort that mattered as much as or more than the economic gains.

The year 1940 was a turning point in Mexican agriculture for reasons more profound than the establishment of CIMMYT. The Mexican revolution of 1910–1920 underlined in blood the old Mexican theme: peasants had been dispossessed of their lands, by Spanish conquistadores, by the church, and then by foreign investors and local political opportunists in the nineteenth century. Although the most genuine peasant movements in the revolution had not won lasting military victory, the governments taking their authority from the new 1917 constitution nonetheless had to promise to carry out the constitution's provisions for massive land reform. After mostly token efforts by Mexican presidents in the 1920s, President Lázaro Cárdenas began in 1934 to push land reform relentlessly, redistributing some eighteen million hectares of land in six years. In 1938, he nationalized the Mexican petroleum industry, which was as important to the United States at that time as it is today. In the process, he made the ruling political party into a much more powerful instrument, gaining for it across all the diverse Mexican regions a kind of enthusiastic support that it had not previously enjoyed. Cárdenas aroused popular expectations and made nationalist sentiment into an effective tool against multinational corporations. He also fashioned a system of political loyalties and patronage that made his party far more frightening—or far more promising, according to one's perspective—than anyone had previously thought possible.

The more conservative forces in Mexico, with strong allies among foreign investors and governments, looked at the presidential elections of 1940 as a crucial test of their power to survive. If the logic of the Cárdenas program were carried further, private capital might have little to say about the future of the country and very limited opportunities for profitable expansion. The newly shaped ruling party could not be opposed, but if it could be won over internally, there would still be a chance to use the new power of the party to achieve a level of national coordination that would make Mexico into a much more efficient and dynamic machine of growth than had seemed possible before Cárdenas's political consolidation. The conservative view did in fact prevail, both in the selection of Manuel Avila Camacho as president and in the political program that he carried forward and left to his successors. The program did not attempt to repudiate the land reform or economic nationalism; rather, it argued that efficiency in production was the key to holding on to the gains made possible by land redistribution and petroleum expropriation. The state would be the patron of economic growth and private investors, not their enemy.

Critics of the Rockefeller-sponsored research program that began in Mexico in 1940 and of the Green Revolution to which it gave birth usually maintain that the basic plan was to design modern agriculture in such a way that it would be available only to highly capitalized large landholders and so would defeat the purpose of land reform. Norman Borlaug, winner of the Nobel prize for the creation of high-yielding wheat varieties at CIMMYT, insists that the Rockefeller Foundation intended to do just the opposite. The idea, says Borlaug, was to develop highly productive agricultural techniques that could be used by relatively poor peasant farmers because high yields could be obtained by intensive use of inputs that would not require mechanization or landholdings of a size large enough to accommodate machines. The program was meant to undergird Mexican agrarian reform, not subvert it.[3]

Borlaug's defense makes a certain sense in the context of the Mexican experience at the beginning of the research program. There is nothing about the new varieties that makes it impossible to use them on small plots, and with proper financial and technical support, peasant producers have made them profitable on a small scale in some regions in Mexico and elsewhere. Furthermore, it would have been politically foolish for the Mexican government to have cooperated in a research program that would be seen immediately as a means of undermining the enormously popular land reform initiative of Cárdenas. It makes sense to believe that the smart money in 1940 would have looked for a way to turn small peasant producers into successful farmers dependent on banks and marketing firms for financing, fertilizers, and pesticides. Of course, over the longer run what actually happened was that unequal access to irrigation, credit, extension services, and political power favored larger producers and left most of the beneficiaries of the land reform in the lurch.

The historical record in the Rockefeller Foundation archives reveals that officers of the foundation and scientific advisers who designed the Rockefeller research program in the early 1940s were to some extent aware of the political and social dangers of the approach they advocated. Recent work by Bruce Jennings of the University of Hawaii in the Rockefeller archives shows that Rockefeller scientists straightforwardly addressed the problem of a "top-down" versus a "bottom-up" strategy: "The plan presented assumes that most rapid progress can be made by starting at the top and expanding downward." They argued that the deficiencies of Mexican farmers, agricultural extension agents, and scientists were such that it would be futile to begin from the bottom. Furthermore, they insisted on the "encouragement, stimulation, and guidance of existing institutions and individuals." As Jennings points out, "Such arguments obscured the sharp cleavages that existed between individuals and institutions. In the aftermath of the revolution and controversies renewed by the Cárdenas administration (1934–1940), was the improvement of agriculture to occur amongst commercial farmers, subsistence cultivators, or the landless?"[4]

In 1941, stimulated by an officer of the Rockefeller Foundation who thought some of these issues bore more serious consideration, the foundation solicited the advice of geographer Carl Sauer of the University of California at Berkeley, a renowned scholar with extensive experience in Mexico. Sauer, in 1941 and in continuing correspondence over several years with people in the foundation, maintained that the problems of Mexican agriculture had far more to do with economic exploitation than with Mexican cultural practices and techniques, which he regarded with admiration. Sauer also believed the technical proposals were dangerous to Mexican peasants and their local economies:

> A good aggressive bunch of American agronomists and plant breeders could ruin the native resources for good and all by pushing their American commercial stocks. . . . And Mexican agriculture cannot be pointed toward standardization on a few commercial types without upsetting native economy and culture hopelessly. The example of Iowa is about the most dangerous of all for Mexico. Unless the Americans understand that, they'd better keep out of this country entirely. This must be approached from an appreciation of native economies as being basically sound.[5]

Sauer continued for many years to propose that the Rockefeller program should proceed by working upward from the problems of the peasant household, patiently solving problems as identified by the peasants themselves. His advice was ignored. So were the warnings of noted botanist Edgar Anderson, whose fieldwork studying the wild ancestors and native varieties of corn laid the base for the Rockefeller program of plant breeding. So were the criticisms of Edmundo Toboada, from 1933 to 1944 the head of the Mexican Office of Experiment Stations, who consistently maintained that research should go forward with the active participation of the peasantry. In Mexico, as Jennings says, "Divisions within the Ministry of Agriculture between American scientists and certain Mexican scientists reflected a basic disagreement about agriculture, central to which were differences regarding the origin and control of science."[6]

Jennings points out that in the mid-1950s, the foundation finally began to pay some attention to the fact that the program threatened to produce, in the words of one of their advisers, "very acute problems with respect to the political control of these benefits. . . . These very benefits may introduce fresh economic disparities with the Mexican economy, which will present political problems not now even dimly perceived by many Mexicans." The response, however, was not to reexamine the goals and methods of the research program, but rather to produce economic and political recommendations for the Mexican government capable of managing "the greatest and most advantageous change in production with the least possible social and economic disruption." By the 1970s, the agricultural economists hired by the foundation to smooth the path of the "disparities" produced by the Green Revolution were able to encapsulate the nature of their work with the straightforward formulation that they were "not interested in theories of how the rich screw the poor."[7]

The real argument of the foundation's program over these several decades from the 1940s to the 1980s is clear enough: any worries over distribution, disruption, employment, or ecological problems may be answered if production gains are consistent and impressive enough. For a time they seemed to have won that argument.

Between 1955 and 1975, per-acre yields of wheat in Mexico increased by a factor of five. Chemical fertilizer use on all crops went from 8,000 tons in 1950 to 346,000 tons in 1965. Pesticides became a standard feature of Mexican agriculture. Several million new acres of land were brought under the plow by the extension of irrigation, forest clearing, and the use of more marginal lands.[8] In spite of an annual population growth rate during most of the post–World War II era of greater than 3 percent, Mexico seemed in the 1960s to be on the verge of permanent food self-sufficiency. During the mid-1960s, grain production had become so successful that the Mexican government found itself turning to subsidies to hold up prices to the farmer while holding down food prices to the consumer. Development analysts believed that low food prices were essential to maintain the momentum of industrial growth, which has averaged an impressive 6 to 7 percent from the end of World War II until the recent recession. Only if the annual 7 percent growth in urban population could be supplied with food at reasonable prices would it be politically possible to maintain downward pressure on wage rates, providing continued incentive for industrial investment. The whole structure of Mexican economic growth was solidly based on the success of the high-yielding varieties and the technical improvements that came with them. Or so it seemed.

To a striking degree, the conservative program begun in 1940 seemed to have won the day by the 1960s. Gains in agricultural productivity provided the base for industrial growth rates matched by few other nations. The doctrine of productive efficiency had proved its value. This point of view, however, was soon to be exposed as dangerous complacency. A conservative English observer and long-time adviser to the Mexican government on agricultural policy points out:

It was disquieting for the nation to find in the 1970s that the country had drifted into the position of becoming an importer of basic foodstuffs; likewise the social conscience had been shocked by revelations of the persisting poverty of many rural people. It had long been assumed that Mexico had sufficient land to feed itself and to provide a substantial quantity of agricultural exports. Technical progress would assure an ever-continuing growth of output and of farmers' incomes. The discovery that all this was no longer inevitable produced a rather violent reaction in public opinion—and a search for remedies.[9]

Clearly one of the reasons for the growth of complacency and the shock that inevitably followed is that the Green Revolution had been such an enormous publicity success, eagerly promoted by the politicians and private interests whose policies and investments received justification from the publicity. In retrospect, it is obvious that the high-yielding varieties were neither so important to the growth of output nor, in themselves, so much to blame for difficulties as most people assumed.

The Mexican agricultural scene is extraordinarily complex. The sheer variety of ecological conditions under which Mexican farmers work is staggering. Regional differences between the arid northwest coast, the humid northeast coast, the temperate altiplano of the central highlands, the tropical Gulf coast, the distinctive regime of high rainfall, high evapotranspiration, and very rapid drainage in the northern Yucatan peninsula, and the windblown southern altiplano of Chiapas are diverse enough; when the literally thousands of combinations of microclimates, topographies, and soil conditions are taken into account, generalization becomes extremely difficult. On top of this physical variety is a cultural variety that is in its own way as complex. No one who has dealt with the Yaquis of the northwest will ever confuse them with the mestizo people of central Mexico or the descendants of the Mayans who predominate in rural Yucatan.

Cultural differences are not incidental to agricultural development, for they influence the way plants are selected, planted, cultivated, and harvested, as well as patterns of land distribution, labor, income, and consumption. To cultural variety must be added a social history that, in spite of the persistence of certain themes of struggle over land tenancy and labor relations, is as different from region to region and even village to village as rainfall patterns or the likely times of frost in mountain valleys. Some areas have been involved in cash crop production since the early days of the Spanish conquest while others remain primarily dedicated to subsistence farming. In some regions, plantations have been the major influence, in some, haciendas, in others, communal landholding, and in still others, small private plots, either owned, rented, sharecropped, or held in usufruct from communities. In some regions, farm families supplement their income by home handicraft production, but in the Puebla Valley, for instance, peasants may continue to farm while moonlighting as workers in automobile assembly lines. Some towns and villages have enjoyed relatively democratic traditions, while others have been dominated by local landowners or political bosses.

It would have been a miracle indeed had the "miracle seeds" provided a national panacea in a country of such diversity. From the scientific point of view, one of the reasons Mexico was such an attractive site for a plant-breeding research program was its extreme ecological diversity, particularly with regard to corn, where dozens of domestic varieties had been evolving along with wild plants for nearly 10,000 years. The adoption of new seed varieties developed at CIMMYT reduced the genetic diversity of the seed stock, but it could not cancel all the ecological diversity that created the original varieties.

New seeds from CIMMYT were adopted by farmers of many different kinds all over Mexico, but certain patterns of adoption can be discerned. Had the patterns been more clearly seen, some of the false expectations might have been avoided. The most spectacular production gains associated with new seed varieties occurred on the northwest coast on relatively large, highly capitalized commercial farms with substantial backing from the Mexican government and foreign investors. The doubling and redoubling of wheat yields that led so many to believe that Mexico's food dependence was over took place primarily in this newly irrigated region. Had farmers been using traditional seed varieties on these virgin soils in the excellent growing conditions of the region, and with the new abundance of water for irrigation, they would have realized substantial production gains in any case. Elsewhere in Mexico, farmers could not afford the necessary fertilizers and machinery, and they did not have reliable access to water. Neither, for these farmers, was wheat the most economically attractive crop given the local dietary patterns and small landholdings that made corn and beans more sensible both for local markets and subsistence. New varieties of corn and beans did find a place among small-holding peasant producers, but with very modest production gains compared to the gains of wheat, cotton, and tomatoes in the northwest. The 6 to 7 percent annual growth in national agricultural production from 1940 to 1965 was caused more by the opening of new lands, the provision of irrigation, land redistribution, and fiscal incentive than it was by new seeds. According to a friendly observer of the Green Revolution in Mexico, a study of Mexican government statistics shows that only one-sixth to one-third of the production gains can be attributed to the seeds themselves and the particular production techniques designed for the new seed stock.[10]

Another factor dampened the success of the new varieties in the late 1960s: government investments in irrigation schemes and the early returns from grains led the large-scale commercial farmers to look for more lucrative crops like tomatoes, melons, and other fruit for export to the United States, oil crops like safflower, livestock and poultry operations based on feed grains, and cotton. While these successful enterprises increased national income, the income was narrowly distributed and consumption of the crops was mostly limited to the relatively well-off urban groups in Mexico and the United States. The goal of food self-sufficiency based on grains receded far into the distance, and hunger and malnutrition still stalked the poor.

While the commercial, large-scale production of the northwest, the Bajio of Guanajuato, La Laguan, and a few other areas found success and basked in the light of a persistent promotional partnership among the government, agricultural researchers, and agribusiness interests, the majority of Mexico's farmers found life considerably less encouraging. Until the early 1970s, credit was hard to come by. Tenancy often remained insecure. Peasants were often located on marginal lands, and even where they were not, the yearly pressure to produce resulted in severe erosion, lowering of water tables, loss of soil fertility, and other environmental problems.

In Los Altos of Oaxaca, for example, peasant farmers contended with the fact that they based their own production on hand-dug wells going down twenty or thirty feet while, nearby, large landowners began to dig much deeper, using motorized drills and pumps. By the mid-1960s, many of the Zapotec and Mixtec people found that a water table that had sustained their people for many centuries was now unreachable. In addition, the larger operations exerted constant pressure to drive small holders off the good valley land onto eroded marginal slopes. Land speculators moved in to develop luxury housing for vacationers and wealthy retirees. The accumulated pressure forced many of these people into the migrant labor stream, where some of them ended up harvesting export tomatoes in Sinaloa or working in garment factories in Los Angeles.

David Ronfeldt of the Rand Corporation has told the moving story of decades of peasant struggle in Atencingo in the state of Puebla. While the American owner of a sugar mill and refinery had been subjected to expropriation under the land reform program and the land turned into a commonly owned *ejido*. (*Ejido* is a landholding form revived by the 1917 constitution and based on traditional Indian patterns, involving individual and shared production, with individual plots held in usufruct.) In spite of the reform, the former owner and the government, interested in foreign exchange earnings from sugar sales, exerted persistent pressure to force the new reform unit into traditional plantation patterns. Although the peasants wanted to farm individual plots growing a variety of subsistence and truck garden crops, they were pushed again and again into maintaining plantation production of sugar to supply the mill. As Ronfeldt says,

> In effect, the government officially licensed monopoly capitalism and exploitation, and a complex agrarian bureaucracy was committed to institutionalization and defense of this system.

Needless to say, the ejiditarios were not in favor of such a system. They preferred an economy and a life-style based on a combination of subsistence agriculture and local market capitalism, with each ejiditario personally controlling a distinct plot of land. Cane was simply not the most profitable or the most useful crop they could cultivate. Rice, tomatoes, and melon often brought higher profits at the market. . . . Moreover, corn and beans met family subsistence needs and brought an easy profit on the market. Parcellization would satisfy the long-standing desires of the peasantry for personal control over their property and livelihood.[11]

While the peasants of Atencingo fought for parcels of land and diversity, in other areas peasants fought for more collective forms of work or control over monocrop production, as in the henequen fields of Yucatan. When Ronfeldt tried to find out whether the story of Atencingo was typical of what was happening to Mexican peasants, he found directly opposite reactions from other students and participants in the struggle for control of the land. Some said that, yes, Atencingo was "the story of Mexico." Others maintained that the case was exceptional. Often the same person would defend both propositions. This is not surprising, for although the general theme of peasant struggle for autonomy and control, usually frustrated, has been constant, the forms that the battle has taken have been as diverse as the cultures and regions of the nation.

Diversity of culture within Mexico was once matched by the genetic diversity of seed stock, but this is no longer the case. Although new CIMMYT seeds were to some extent developed for Mexican conditions, they could not possibly be well adapted to all of Mexico's diversity of ecological conditions. The new corn seeds, for example, were more successful in the American Midwest than they were in Mexico. While corn yields did increase in some Mexican regions, the loss of genetic diversity in seed stocks and particular characteristics that allowed for greater nutrient uptake required heavier and more nearly universal use of fertilizers and pesticides and made secure access to irrigation water much more important. The loss of genetic diversity in corn threatens the ecological sustainability of agriculture everywhere in the world, not just in Mexico; it means that the way to an agriculture with fewer pesticides and more modest water and fertilizer requirements will be more difficult to find.

Anyone who has visited Mexico City will be painfully aware of another feature of the kind of development that has taken place, and for which capital intensive agriculture must bear some portion of responsibility. The fifteen million people in this city (or nine million or twenty, according to where one sets boundaries to the metropolitan area) live with massive housing problems, water shortages, severe air and water pollution, transportation snags, and relentless growth. Other cities of Mexico, expecially the border towns like Juarez and Tijuana, have even more rapid growth rates than the capital and problems as severe in many respects. Tijuana, for instance, has a serious shortage of drinking water with no foreseeable solution to it, although the city will almost certainly continue to grow at something between 7 and 10 percent per year until some absolute limit of tolerance is reached. While the 6 to 7 percent growth rate of Mexican industry has nearly matched the urbanization rate, employment has not kept pace. American analysts commonly recommend that Mexico resolve its indebtedness by a cut in social expenditures of the type that cushion the effects of rapid urbanization and unemployment; it is difficult to believe that such people have visited Mexican slums or that they have any sense of ordinary decency left.

Of course, for all of this, petroleum was to be the solution. The nationalized petroleum company (PEMEX) has offered fertilizers at well below world market prices for many years, sustaining growth in agriculture. Pesticide feed stock chemicals are available from PEMEX cheaply as well. Road building has been subsidized by PEMEX for forty years. Oil income could meet social expenditures, buy what imports must be bought, and provide investment toward more permanent solutions. It is precisely the failure of petroleum to work these miracles that points up the most fundamental problem of the pattern of agricultural development followed in Mexico. The strikingly uneven pattern of growth and distribution among sectors of the Mexican rural and urban economy continues to plague the country with inflation, unemployment, and instability.

Twenty years ago, economists spoke of "dual economies" as one of the central features of underdevelopment. According to the theory, the basic problem was that there were typically two economies in poor countries: one was dynamic and well integrated with national and international markets, the other was stagnant because it was largely based on subsistence and was not integrated into markets. The dynamic section of the economy was responsive to policy initiatives by government and to entrepreneurial initiative, while the stagnant section was unresponsive. The stagnant subsistence economy was, in this view, an obstacle to development. Successful development policies would increasingly integrate the second economy with the first. This analysis has been largely discredited in recent years because it has been shown that the supposedly stagnant economy was in a sense a creation of a certain pattern of development, fulfilling purposes vital to the continued function of the dynamic sector.

This latter-day view, in a general way, amounts to a fair description of the Mexican rural scene. The majority of poor farming and rural labor families produces most of the nation's food supplies, while agrarian reform has kept enough of these people in the countryside to hold down urbanization rates with little direct investment required. Land reform was taken just far enough to make it possible to continue to bleed the countryside without causing a fatal hemorrhage. At the same time, public investment in irrigation and agriculture research has gone mostly to the more concentrated commercial sector, which is able to acquire migrant labor from the poverty-striken sector when necessary and able to choose among crops and technologies to maximize profits. The dynamic sector concerned itself with basic foodstuffs only so long as it was lucrative to do so, although government investment in this sector was justified as a means of increasing basic food production. Now, the commercial farms have left that task largely to others.

The majority of poor farmers have shown again and again that they are responsive to entrepreneurial initiative and opportunities as well as to government support. They work hard, they take risks, and they participate in the market economy—all that is stagnant about their share of the economy is their income. As desperate farmers try to overcome their poverty, they do serious damage to the land because they cannot afford fallowing, erosion control, forest protection, or proper fertilization techniques. They cannot afford to leave marginal, sensitive land for more fertile, durable soils. The sharp division between a dynamic commercial sector skimming the cream off the rural economy while poor peasants produce most of the food under worsening conditions describes the situation in much of Latin America, Asia, and Africa where similar techniques have been used. Some suggest Poland's rural development mimics Mexico's closely, where research has been directed toward the success of capital-intensive state farms at the expense of a larger sector of independent and cooperatively organized peasant producers.[12]

For Mexico, it would be unfair to blame this situation entirely on CIMMYT, agricultural research, or on Green Revolution technology. The very complexity of Mexican geography, culture, and politics makes it absurd to do so. It would also be absurd to say that Mexican farmers would be better off without a modern research establishment—the problems of Mexican agriculture must to some extent be solved by improved technologies. No one can pretend, in the face of such immense complexity, to have any single answer to the problems. That is the point.

One way in which agricultural research went wrong was precisely in saying and allowing it to be said that some miracle was being produced. Neither miracle seeds nor miracle petroleum can solve such complicated human problems. The task of development is more social and political than it is technological, as hard as that may be for many people trained in American universities to believe. Neither infusions of new productivity (through seed varieties, for example), nor infusions of cash (through petroleum) can resolve the really difficult questions of poverty, environmental protection, cultural integrity, and political sovereignty. After all, Mexico has provided gold and silver to the world in massive quantities since the sixteenth century. It is precisely the production of easy wealth by small sectors of the economy that has created the top-heavy, highly stratified, corruption-ridden society in question. Miraculous sources of income will never be the salvation of Mexico—they are more likely a curse.

The program of technological development in agriculture in which Americans participate in Mexico has been, from 1940 to the present, a political program, promising one particular kind of answer to economic problems. No such major change in agricultural technique can be anything other than a political program. There are strong reasons to believe that this has been the wrong program, technically and politically. Peasant unions have protested against it. Students and professors at Mexico's national agricultural college in Chapingo have gone on strike en masse to insist on another kind of program more suited to small-scale, poor farmers and to the diversity of Mexican agriculture. Dozens of Mexican scholars, and a few outside of Mexico, have rejected it as a sensible solution to the major problems of rural Mexico. And yet, the program goes on.

Norman Borlaug insists that for all its faults, the program has bought time, time in which socioeconomic changes may occur.[13] This is, at best, a point of view filled with that dangerous sort of innocence we spoke of earlier. For the program has not simply provided for increased production in a neutral fashion while a more permanent solution is worked out. Rather, it has reinforced the very problems of stratification and distribution that are the main problems, while promoting a set of delusions about miraculous results that simply cannot come true.

It would be sad enough if all this had to happen to educate the American research establishment about the social implications of technological development. Unfortunately, the most important lessons apparently have not been learned—although slowly and after intense criticism, some of the narrower ecological implications of the new agricultural technologies have been accepted by American researchers. Nonetheless, they still seem to be incapable of understanding ecology in its larger meaning, one that includes human society in the ecological equation.

Perhaps we need to look at the social ecology of the American land-grant university system and the agricultural research establishment. There is a distinctive ambience about the places where agricultural education and research go on, and I fear, it may be an environment in which certain kinds of messages are systematically lost.

The atmosphere of the land-grant university town is almost always curiously contradictory. On the one hand, such towns as Manhattan, Kansas, or Davis, California, are deeply provincial, reflective of the close ties they maintain with the surrounding rural communities. The majority of students and professors come from farming or small-town backgrounds and share the largely conservative values of a generally prosperous Anglo-Saxon culture.

On the other hand, the land-grant university has a strongly cosmopolitan flair. Its student bodies typically include sizeable groups from India, Pakistan, Israel, Latin America, Africa, and the West Indies. Some of these foreign students take professorships or stay on at research or extension tasks, introducing into American rural communities an element of foreign aid in reverse, which tends to bemuse when it does not anger farm families. American professors who know nothing of Chicago, New York, or San Francisco often have spent time in rural areas outside of New Delhi, Brasília, Medellín, Lagos, or Tel Aviv. Agriculture professors often speak a foreign language more fluently and more colloquially than their colleagues in literature and foreign language departments, and they may have friends scattered across two or three continents. They often pride themselves on knowing the "real people" of the countries in which they have worked, in contrast to their colleagues in social sciences and humanities who, if they are anthropologists, know people from tiny remnant indigenous groups or, if they are from other disciplines, know scholars and politicians. They tend to believe that their experience has taught them at least one important lesson—that the real people in all countries are faced with similar problems, which are to be solved largely by technological means. They tend to feel that barriers to this sort of problem solving are unnecessarily thrown up by politicians, dreamers, agitators, and fellow scientists who have become lost in the vagaries of social and economic theory. They tend not to see that their own work is itself a form of political activity. They deny that their work carries the moral responsibilities inherent in political action.

Another way of describing this situation is to say that people who are provincial and empirical by inclination have received validation for their views from the fact that their skills and knowledge are highly sought after by people like themselves all over the world. The fact that American agriculture itself is dominant in world markets confirms the fundamental rightness of such a view.

Yet a dark cloud hangs over the landscape of American agricultural universities: hunger, malnutrition, rural unemployment, and rural violence remain intractable in almost every country in which these new agri-industrial techniques have been adopted on a large scale under American and European tutelage. It is commonly and easily said that the modernization of agriculture has at least kept a bad situation from becoming worse, but there is increasing suspicion even at the land-grant schools that the answer is much more difficult than that. In spite of all the departments of rural sociology and organizational behavior, the themes of technical research remain separate from the themes of what matters and what works in a social context. No number of books and monographs on the failures of new agricultural techniques to work miracles, however much they may be read, will actually succeed in seriously influencing the direction of agricultural research so long as there is no reexamination of absolutely fundamental assumptions. Social scientists and ecologists cannot be expected to come in toward the end of a program of technological development and simply fix what has gone wrong, as the Rockefeller Foundation asked its social science team to do in Mexico in the 1950s. Some actions are so far reaching in their effects that a failure to accept responsibility from the beginning virtually precludes the possibility of responsible action later.

Agricultural researchers have for too long lived in an international subculture, a subculture that exists at CIMMYT as well as in Davis, California, or Manhattan, Kansas. As in Mexico, their isolation is maintained by governments and foundations with agendas of their own. In recent years, there has been a strong current of unease and even protest at some landgrant universities, as students and professors have begun to face the ways in which their work and its applications have been shaped by the rather narrow purposes of those who provide the funds. It is essential to realize, however, that it is not simply a question of funds provided to public research to produce private profits, as in the well-known case of pesticides research and chemical companies. Much more important is the fact that, with every intention of doing well by humankind, those who have patronized the research have certain assumptions about what is good for the human race, a certain innocence protected from doubt by power and influence.

I do not imagine that the agricultural economists at CIMMYT will soon become interested in "theories about how the rich screw the poor." Perhaps a better place to begin is where Carl Sauer began forty years ago—by suggesting that the problems appearing in American agriculture because of techniques that are highly energy, chemical, and capital intensive gave warning that such techniques should not be used as a model for agriculture research in other countries. The failure to adopt Sauer's kind of humility about American agriculture has led American researchers to act grandly and arrogantly, as well as innocently, abroad. At the same time, we find the agricultural research establishment in the United States glossing over its domestic failures by citing the international success of miracle seeds. Historically, science and technology made their first advances by rejecting the idea of miracles in the natural world. Perhaps it would be best to return to that position.

12

The Practice of Stewardship

John Todd

It will come as no news that the world's ecological fabric is under siege. The news is in the magnitude of the destruction. Today in the United States, desertification, or the loss of soil as a result of contemporary agricultural practices, is occurring over a greater land area than was encompassed by the original thirteen colonies. It is worse than the Dust Bowl of the 1930s. Globally, the crisis is amplified, and deforestation, the loss of natural resources, has already occurred over a land mass larger than all of Brazil. Richer countries, while trying to save their own, are purchasing the forests of poor nations, particularly in the tropics. As a consequence, biological inequities are rapidly growing among peoples and regions of the world.

There are a number of counters to this unhappy trend, and one of the most important of these can be found in many of the New Age communities that are predicated on the spiritual and practical notions and ethical dimensions of stewardship. This approach is important, and I predict that the seeds of the next agricultural revolution are in the gardening and farming in such communities. The land is seen as sacred and is honored. In this context, agriculture and agricultural sciences are very different—they are to a large degree based upon finding analogs of the complex workings of ecology. It is farming in the image of nature and not machines or new genes that will transform agriculture. Inherent in this attitude is a larger task, namely, of ecological healing. If such communities are to flower and succeed, then their work should expand beyond their gardens and involve no less than the restoration of the planet.

For the past thirteen years I have been involved in a search for a science of earth stewardship. One aspect of this search has been to travel to places where nature is especially bountiful and find out why. I would try to assemble the constellation of characteristics of a place—its geological, topographical, ecological, and historical elements—and weave them into a "picture" that I could see. Many of these places, while being ecologically different from each other, had patterns in common. It is these patterns that provide the clues to the practice of stewardship.

The same approach can be applied to the study of regions where humans have lived in harmony with nature. There are a few examples of this harmony and bounty being sustained by humans for millenia. In parts of Indonesia, particularly Bali, agriculture is still practiced as it has been for a thousand years; fortunately my guide to these communities was the extraordinary anthropologist Margaret Mead. She took my wife, Nancy Jack Todd, and me to many of the areas that she and Gregory Bateson had worked within in the 1930s. In discussing villages and settlements with her, we began to see the continuum between the mindscape and the landscape, between agriculture and art. The land was within and without, with little discontinuity between the various elements. Bali inspired my thinking about settlements and villages.

But the farm I want to describe was one I visited near Bandung in central Java. On one small landscape, cultivated continuously for centuries, was a farm that reflected in miniature the major restorative processes in nature. Most agricultures are short lived, a few centuries at the most, before the land tires and falls into disrepair. Here I was looking at a farm where the fertility was probably increasing each year and had for hundreds of years. It was an example of a true partnership between people and the land.

I began to see some of the patterns I had observed in the wild places. What I discovered is a delicate balance among the kinds of agricultures; all the major types are interwoven on one piece of land. Trees, livestock, grains, grasses, vegetables, and fish are all present and no one of these is allowed to dominate. Equally significant are the relationships between the water and aquaculture and between the land and agriculture. They are worked together. In the West we almost never join water and land this way and that may explain why our agricultures are relatively short lived.

The farm on Java is hilly, and although the native forest is gone, it has been replaced by a domestic "forest" of trees with economic and food value. The domestic "forest" first and foremost protects the hillsides, the farm houses and buildings, and the agriculture and fish culture below.

Water enters the farm in a relatively pure state via an aqueduct or ditch on the contour of the land. To charge it with nutrients so it can fertilize as well as irrigate the crops, the aqueduct is passed directly under the animal sheds and the household latrine. The manure enriches the water, which is subsequently aerated by passing over a small waterfall. At this point the enriched water flows between deep channels amongst raised bed crops. The water does not splash directly onto the crops, but seeps laterally into the raised beds. In this ingenious way, animal and human wastes are used while minimizing contamination of crops by pathogens harmful to animals or humans.

The gardens filter and to a degree purify the water. Water not absorbed or lost to the garden joins in a channel and flows into small ponds where fish are hatched and raised. They require the high-quality water that comes from the vegetable garden. The banks of the aquaculture ponds are planted with a variety of tuberous plants; the leaves are fed to the fish and the tubers go to the livestock.

The presence of fish in the ponds causes the water to become loaded with fish wastes, so it is enriched for a second time and flows into rice paddies that it both floods and fertilizes. Again, the nutrient and purification cycle is repeated. The rice filters out the nutrients and organic materials, and water leaving the paddy is again in a somewhat purer state. Finally, the water flows to the bottom reaches of the farm where it enters a communal and partially managed large pond. From time to time, organic matter, including sediments, is taken from the pond and added to the soils on the higher reaches of the landscape.

This type of farming is a good, albeit rare, example of millenium-old agriculture. What is most interesting about the farm is that the farmers got the relationships right, and the patterns of balanced interdependence between the various components. Obviously, if pesticides were applied to that farm, the fish, highly sensitive to toxins, would die, and the chain of ecological relationships would be broken. The ecological integrity of the farm would break down, as is the case on farm after farm throughout the world.

For the student of planetary healing, another interesting approach is to study places that are ecologically reduced. Almost invariably, one finds the earth has lost its ability to hold moisture for long periods. The soil is too porous, too compacted. This is the condition on a goodly portion of the planet, which has been overexploited by humans and their livestock.

When soils lose their spongy water-absorbing qualities, flash floods, drought, and loss of topsoils become the norm. Lack of fresh water near the soil surface is the major problem in the restorative process. The challenge for a science of earth stewardship would be to find ways to reverse the process, to learn how porous soils could be made to hold water again.

I was confronted with just such a challenge on a coral island in the middle of the Indian Ocean. Several years ago, the Threshold Foundation, based in London, asked several colleagues and me to visit an atoll in the Seychelles. As part of their fledgling "Islands in the Sun Project," we were to investigate the possibility of ecologically diversifying the small atoll. Now a major problem on coral islands is the lack of fresh water. Coralline soils are notorious for their inability to hold water and for the absence of surface lakes or ponds. Water is normally obtained from household cisterns and by pumping from fresh-water underground lenses (underground water reservoirs), which are often found under coral islands. Unfortunately, these lenses are readily used up, and salt water invades. This is what was happening on the atoll in the Seychelles.

While reflecting on the inability of the soils to hold water at the surface, I remembered a research paper I had read about a decade earlier. Several Russian scientists had wanted to find out why bogs and small ponds were often formed on top of rubble-heap hills that under normal conditions wouldn't hold water for long. They discovered a biological process, which occurs over long periods of time, that they called gley formation. Gley is like a biological sealant or plastic. It forms in the absence of oxygen and when the carbon-to-nitrogen ratio of the accumulated plant material is just right.

I was curious to find out if surface lakes could be made on coral islands through speeded up or "quick-time" gley formation. The atoll provided a chance to experiment, to simulate a natural process specific at least to bogs in Russia. A large pond was dug with a small backhoe. Coconuts were shredded and placed six inches thick on the pond bottom. To add nitrogen and other essential ingredients, we gathered and put in place a six-inch layer of shredded wild papaya trunks, leaves, and stems. Papaya was a prominent understory plant and suited to our task. Then, to drive out the oxygen and produce an anaerobic environment, we laid down a layer of flimsy hole-filled plastic sheet and on top of that six inches of coral sand. The bottom and sides were tamped down.

The pond was given an initial infusion of well water and then the rainy season came. The pond filled up and stayed that way. The experiment worked, and the pond began to act as an ecological magnet for all kinds of life. The gardens and livestock were diversified. Sir Peter Scott discovered that the fresh-water pond was attracting migratory birds that rarely, if ever, land on islands in the vast expanse of the Indian Ocean.

There is another side to the tale. The outstanding soil biologist Stuart Hill, of McGill University, was with us. He lamented that the soils were so alkaline that few domestic plants, like vegetables, could grow in them. So, he decided to neutralize the garden soils by adding acids from a readily available biological source. Knowing that compost, just before the process is completed, releases organic acids, he made a compost heap; at the acid-releasing stage, he placed the compost, not yet ready by ordinary standards, on the gardens so they could release their acids and make the soils habitable for human foods. He had found an ecological equivalent to the expensive petrochemical-based methods.

These stories illustrate that there is a vast storehouse of knowledge currently locked up in the insular reaches of academic and scientific institutions, which can be used to form a science and practice of earth stewardship. Inherent in the workings of nature are the keys to the restorative process.

Another dimension to this is quite mysterious and wonderful in its implications. Here I am writing about nature's metapatterns on a global scale. I had been schooled in the writings of the GAIAN (from *Gaia*, the Greek goddess of the earth) of G. Evelyn Hutchinson, Alfred Redfield, and more recently, James Lovelock and Lynn Margulis. They had, through their researches, begun to identify the earth as an organism or complete being with its own global mechanisms for self-regulation. They were suggesting, much as had the ancients, that the earth is alive. One implication of this came to me while talking to the poet Gary Snyder, and it has shaped my life since. Gary and I were discussing our love for the Mediterranean world, the light and sense of time and space connected through the vast panorama of Western civilization. Then we began to lament its damaged ecology. The Mediterranean is a dying sea with its load of industrial pollution and its diminished gene pools, especially in the marshes that are the nurseries of so much marine life. The destruction of salt marshes began with the Romans, reached a frenzy under Mussolini in the 1930s, and has continued elsewhere in the region ever since. The destroyers were, of course, trying to rid the area of the malaria-carrying mosquito, but in doing so by means of the destruction of whole ecosystems, the source of much marine life was lost. Such reckless methods are no longer needed; for example, biologists recently have found a strain of bacteria, *Bacillus thuringensis*, that is specifically lethal to mosquitoes. This mosquito-killing strain of bacteria doesn't kill other kinds of aquatic life. What this discovery demonstrates is that ecological means to control a major problem like malaria can be found if one looks long and hard enough.

Gary Snyder and I began to discuss what the ancient ecology of the Mediterranean might have been like. We know from early writings that many of the Greek islands, now devoid of most trees and soil and without much fresh water, were once richly forested and dotted with springs. The Mediterranean lands have lost much of the ancient ecological bounty. When Gary and I started talking, we both assumed that the Mediterranean we longed to know and see was no more. Then in a moment of shared insight we realized that this might not be true, for the ancient ecology might just exist scattered in bits and pieces in various parts of the world waiting to be reassembled. The earth has analogous Mediterranean types of environments. Not the same species exactly, but analogous forms with structural relationships that are comparable. Mediterranean-like places in California, Chile, Australia, bits of Africa, and India have in aggregate the vast array of species needed to recreate the ancient ecological integrity of the Mediterranean region. Modern-day stewards could carefully assemble a "bouillabaisse" of living forms on an island, and if exploitation were prevented, the ancient Mediterranean would reappear. The same patterns and diversity would flourish again. Out of that conversation I began to glimpse a deeper meaning for the words *stewardship* and *planetary healing*.

Those reflections prompted the founding of two new organizations, The Company of Stewards, Inc., and a research and communication organization, Ocean Arks International, to explore various facets of planetary stewardship. Twelve years with the New Alchemy organization had taught us ecological design. Now we wanted to apply what we had learned in a variety of regions, including the Mediterranean.

I would like to carry out a restoration experiment on a Mediterranean island or coast that is semiarid and ecologically impoverished. For the past couple of years I have been working on the technologies and ecological strategies needed to accomplish, in a relatively short period of time, the restoration. One of my first tasks would be to make salt marshes in low-lying valleys. Using New Alchemy-type sail-wing windmills, we would pump sea water into the low-lying coastal valleys. The sea water would flow by gravity back to the sea. The windmills would provide a technologically created analog of tidal action. The new salt marshes would then be planted with a wide variety of organisms and seeded with marine creatures collected from relic marshes still left in the Mediterranean. Ecologically based mariculture would be linked to the project at this juncture and would provide the restoration process with an economic base.

Along the edges of the salt marsh, salt-tolerant plants, like the carob tree, would be planted. These plants have the ability to separate the salt from the water in their tissues. Slightly higher up on the banks, fresh water would be available, to a modest degree, to salt-sensitive plants. The salt marshes would also act as catch basins for seasonal rains. They would become increasingly brackish and able to support diversified life forms. In the sea nearby, marine life would be enhanced. It would be the beginning of an ecologically complex restorative cycle.

Another strategy I have been developing is more technological and is particularly useful in impoverished, arid areas. With the aid of special greenhouses or bioshelters, sea water would be passively distilled to nurture young forests. The plan would work as follows and is based on structures developed and experiments undertaken at New Alchemy and in my own house. A dozen or so bioshelters, approximately fifty feet in diameter and built like the pillow-dome structure at New Alchemy, would be pitched in a circle like an Indian encampment. Inside the central zone of each structure would be clusters of translucent solar silos or solar-algae ponds such as those we currently use to grow fish and heat and cool buildings. During the day relatively cool sea water would be pumped into the ponds. The temperature differential between the ponds and the air would be enough to cause the tanks to sweat fresh water down their sides onto the ground. At night the air would lose its heat to the atmosphere, and the moisture-laden air within would attach to the inside of the bioshelter skin and "rain" down onto the ground around the periphery of the building. In these wet zones caused by the "weeping" of the bioshelter, trees and other plants would be planted. After their roots were deeply established and compost soils prepared, the protective embryo of the bioshelter would be lifted off and taken to a new site to repeat the process. The newly liberated ecosystem would be left behind.

Hardy trees would be planted outside adjacent to the bioshelters to further diversify the restoration process. Each bioshelter might be in place for two or three years before being moved to the next locale. During this period they would do double duty as maricultural facilities and salt producers.

There are many variations of the salt marsh and bioshelter schemes and a number of intermediate approaches. Taken together, they add up to a veritable assembly of biotechnologies that can serve the restoration process.

For some time I have been aware of the power of symbols on the imagination—for example, how Concorde jetliners or space colonies influence the mindscapes of the future. The arks or bioshelters we created at New Alchemy were intended both as tools and as symbols of a biological and solar age. They were to express qualities of immanence and stewardship, much in the way the space ship expresses transcendental and industrial qualities. For many years I wanted to combine the vision of restoring the lands with protecting the seas and informing the earth's stewards. In my mind it came to be a great sailing ship, powered by the wind and sun and filled with live organisms for ecological healing. This biological ship of hope would be a global voyager serving people who desire biological resources and the requisite skills to heal their lands.

So, Ocean Arks International began to develop such a ship, to be named the *Margaret Mead*. Its decks and cabin roof are to be made of wave-resistant and light-transmitting materials. Within, hundreds of thousands of tree seedlings could be grown under guaranteed conditions, and millions of young marine fish could be transported live on board to coastal fish farmers. It would also house a library, tools, and staff capable of undertaking really challenging stewardship experiments.

A few years ago we built a one-fifth scale, fifty-foot model to research new sailing rigs. In the process I learned enough to know that naval architecture and marine engineering hadn't evolved far enough to allow me to undertake building the ship I had in mind. As a consequence, Ocean Arks International has begun to develop small, advanced-design sail-powered fishing and transport vessels that might be useful to the millions of Third World fishermen currently plagued with nonexistent spare parts, lack of information, and erratic and expensive sources of fuel. We have called these fast, multihull vessels "Ocean Pickups." Before long they will be sailing and fishing in a variety of fishing communities.

In the meantime, I am slowly assembling the skills to develop "Ocean Arks." They may not be as big as originally conceived, and in fact, their strength, beauty, and usefulness may be enhanced by their numbers rather than by their size. My dream is that within the next few years planetary stewards will be able to travel from Findhorn to Auroville in India, and Auroville to Chinook Community in the northwest of North America, and on to New Alchemy and Ocean Arks International. I should like the Ocean Arks to be fast and safe and works of great beauty so that stewardship will have workhorses that feed both humanity and the individual imagination. They will allow us to gather the collective land-tending skills of many cultures and link them together for application. Such gathering and linking will be the work of generations and wonderful stuff, too.

13 An Agroecological Approach to Sustainable Agriculture

Stephen R. Gliessman

Agriculture in the United States has enjoyed several decades of increasing yields and burgeoning surpluses. But there are many signs that these increases are leveling off or are even in danger of dropping. Soil erosion rates have reached alarming levels in some regions of the country. Many pesticides are becoming less effective as the pests build up resistance to the sprays. Crop systems based on costly inputs of energy derived from a diminishing supply of fossil fuel are damaging the environment as well as the pocketbook. Scientists as well as agriculturalists are beginning to question the ability of many large-scale agricultural practices to sustain present production levels. Conventional agricultural researchers are seeking solutions, and many find immediate means of maintaining yields, but basic questions are not being addressed. How can agriculture be more in balance with the natural environment? How can agriculture depend less

For this chapter, ideas, support, suggestions, and a fair share of criticism have been gratefully received from the following people: Alba Gonzalez Jacome, Roberto Garcia Espinosa, Moises Amador Alarcon, John Vandermeer, Miguel Altieri, Doug Boucher, John Lambert, and many others. I especially appreciate the excellent editing by Kay Thornley and layout and typing by Kima Muiretta.

on costly inputs and still produce enough food and fiber to supply existing demands? An agroecological focus in research addresses such questions and, in the process, establishes a framework for long-term sustainability of agricultural systems.

Agricultural ecology—agroecology—is based on the premise that the short-term, mainly economic focus of food production must be redirected toward long-term management systems—systems based on cycles and interactions found in natural systems. The term *agroecology* is relatively new, yet the practice is as old as cultivation itself. Past civilizations often modeled their farms after the natural environment. And though rooted in folklore or intuition, many indigenous or traditional farming practices have a scientific basis that is only beginning to be understood.[1]

Recently, researchers in many parts of the world have begun to merge the science of ecology with selected practices of farming past and present.[2] In taking on the characteristics of both its parents, agroecology has become applied ecology and systematic farming. An agroecological focus goes beyond producing crops; it delves deeply into the complex of factors that make up an agricultural system. Great importance is placed upon the internal flow of energy and recycling of nutrients. It seeks to define the role of a wide variety of plants and animals, either singly or in mixtures, in maintaining the long-term balance of the system. Much of the knowledge that ecologists have gained about the factors and processes of natural systems, such as forests or meadows, can be applied to agriculture.

An ecosystem concept is basic to ecology and serves as an important means of defining a unit of study. The boundaries of the ecosystem depend on the scope of the study involved, varying from something as small as the life inside a flower to something as large as an entire watershed. It includes not only the physical components such as the soil, water, air, and light in the system but also all of the living organisms present, their interactions with each other, and their responses to the physical factors around them. Ecologists have used the concept of an ecosystem to study inputs and outputs of a system, as well as how materials move around within the system. By comparing ecosystems, the effects of different disturbances, caused by either nature or humans can be estimated in terms of how they affect the natural balance of resource use. This same concept, the agricultural ecosystem or agroecosystem, has recently emerged in studies of agriculture in an attempt to integrate the great diversity of factors affecting farm systems.[3]

Modeled after the diversity of plants in a natural system, traditional mixed cropping (polyculture) systems still feed most people of many tropical developing countries. Such systems are well adapted to use by farmers with limited access to economic resources and have evolved in response to the particular combination of cultural and environmental conditions of a region, often over long periods of time. In the lowland region of Tabasco, Mexico, for example, evidence indicates that such cropping systems have been practiced more or less continuously since early Mayan time.[4]

In tropical regions, cropping systems that include mixed or multiple crop arrangements have been found to be more productive than those managed with single crop patterns.[5] These agroecosystems intensify production and permit better use of resources by planting more crops in the same space and shortening the time between cropping cycles. Although such cropping systems are very old, little is known about how well they work or why. Several of my colleagues and I have been studying the cropping systems commonly used in Tabasco, gathering information about the polyculture planting of corn, beans, and squash. Examining the results of some of these studies illustrates an agroecological approach to researching agriculture.

In one case, a system of corn, beans, and squash was studied in an area where local farmers had long employed this combination using traditional methods and varieties of the region.[6] Two years previously the area had been planted with corn and was then allowed to lie fallow until the site was chosen for study. Using a machete for the initial clearing, a planting surface with a deep mulch cover was prepared. Local farmers advised against fire, stating it was used only on plots that had been fallowed for many years; if fire were used any sooner, weeds would quickly invade and interfere with the crop development. In the polyculture research plots, seeds were dropped into holes made with a planting stick: five seeds of corn and four seeds of beans dropped in one hole, together, every pace (about one meter apart), and three seeds of squash in holes made separately every three meters. The corn was a local three-month variety, the bean a local climbing variety of *Vigna*, and the squash a local round-fruited variety. For comparison, monoculture plots of several densities were planted for all three crops, both above and below the density used in the polyculture. Only locally used cultural practices were applied, including a weeding with machetes thirty and sixty days after planting, and the doubling over of the corn several weeks before final harvest. Some plots were left unweeded to observe the ability of the different densities to yield well with weeds present.

In a comparison of the yields of each crop at each of the monocultural densities with the polyculture, both with and without weeding, the highest corn yield was obtained in the weeded polyculture (Table 1). Of the monocultures, the high-density planting achieved the highest yields. Bean yields

	Monoculture Densities				Poly-culture
	Very Low	Low	High	Very High	
Densities of corn	33,300	40,000	66,600	100,000	50,000
Yield (kg/ha) with weeding	990.0	1,150.0	1,230.0	1,170.0	1,720.0
Yield (kg/ha) unweeded	846.5	785.8	1,492.6	1,108.4	1,285.6
Densities of beans	56,800	64,000	100,000	133,200	40,000
Yield (kg/ha) with weeding	425	740	610	695	110
Yield (kg/ha) unweeded	140	122	344	438	75
Densities of squash	1,200	1,875	7,500	30,000	3,330
Yield (kg/ha) with weeding	15	250	430	225	80
Yield (kg/ha) unweeded	177	95	143	99	9.6

TABLE 1. Yields of the Polyculture of Corn, Beans, and Squash Compared with Monocultures Planted at Four Different Densities, C-34 site, Cardenas, Tabasco, Mexico
Note: Densities expressed as number of plants per hectare (ha).
Source: Data from Amador, "Comportamiento de tres especies" (note 6).

were reduced considerably in the polyculture compared to monoculture yields; yields were uniformly reduced at all densities without weeding. Squash followed the same trend, although the yield reduction in polyculture was more notable. In general, for all three crops, the best yields were obtained with weeding, although the still significant yields without weeding demonstrate the resistance such cropping systems have to weed presence.

Another indication of system productivity is the total vegetative material (or biomass) accumulated by the end of the growing season. Since much of this material is recycled into the system after harvest, serving as an input of nutrients and organic matter, biomass can be linked to long-term productivity. As Table 2 shows, biomass production followed the same tendencies observed for crop yields. For corn, the highest total biomass was achieved in the polyculture. For beans, the lowest biomass production

	Monoculture Densities				Poly-culture
	Very Low	Low	High	Very High	
Densities of corn					
Biomass dry matter (kg/ha)	2,800	3,150	4,450	4,950	5,900
Densities of beans					
Biomass dry matter (kg/ha)	800	850	775	1,350	300
Densities of squash					
Biomass dry matter (kg/ha)	250	950	1,250	800	450

TABLE 2. Biomass Production in Weeded Plots of the Polyculture of Corn, Beans, and Squash Compared with Monocultures Planted at Four Different Densities, C-34 site, Cardenas, Tabasco, Mexico

Note: Densities expressed as number of plants per hectare (ha).

Source: Data from Amador, "Comportamiento de tres especies" (note 6).

was achieved with the polyculture; for squash, the polyculture was more intermediate.

When the crop yields and biomass are added together for the three crop components of the polyculture, we see how the polyculture could yield more than the monocultures regardless of their density. The polyculture produced close to seven tons of biomass dry matter per hectare plus the harvest of three different crops from the same piece of land. In such a polyculture, corn is apparently the most important crop in terms of satisfying the food needs of the local inhabitants. Thus, increased corn yield makes it very advantageous to plant the mixture, which has the advantage of producing beans and squash at the same time.

An analysis of nutrient distribution in a polyculture cropping system compared to the monocrop can tell us a lot about nutrient dynamics and cycling in an agroecosystem. When nitrogen was analyzed in the vegetative part of the crop system before and after harvest and then adjusted for the amount of nitrogen removed with the harvested material, it was found that the polyculture system had a net gain in the nitrogen balance, whereas a net loss was experienced in the monoculture (Table 3). This difference is most likely the result of biologically fixed nitrogen put into the system through the activity of specific bacteria (*Rhizobium* spp.) in the roots of the bean plants. Hence, the mutual relationships of corn, beans, and squash are of benefit not only to the organisms involved but also to the entire cropping system. Such beneficial relationships directly increase crop yields (Table 1) and improve resource availability in the long term (Table 3).

Given these results, our research then focused on the mechanisms that might be responsible for such yield increases. Boucher and Espinosa found that the polyculture cropping pattern was capable of considerably reducing the depletion rates of nitrogen from the soil. This reduced depletion may be due to increased nodulation and bacterial activity in the roots of the bean plants when interplanted at certain spacings with corn.[7] How the stimulation of such activity takes place is the subject of further research.

Another interesting aspect of the dynamics of the corn, beans, and squash polyculture system is the frequency with which we observed a reduction in pest damage in the mixed plantings.[8] Studies have shown that this reduction occurs through a combination of effects such as increased populations of beneficial predators and parasites of the pest insects or through the modification of the habitat in such a way that makes it more difficult for the pest to find the crop it prefers.[9] By increasing the diversity of the system, either by putting more crop species in or by changing crops more frequently, it becomes difficult for the pest insects to reach population sizes large enough to cause serious damage.

Some pest management practices at first may seem like simple folklore,

System	Initial (A)	Nitrogen Content (kg/ha) in Biomass		
		Final (B)	Yield Loss (C)	Balance (B − A) + C
Corn/bean polyculture	138.6	109.4	69.4	+40.2
Corn monoculture	260.2	192.2	49.2	−18.8
Crop species:				
Corn: *Zea mays*				
Beans: *Phaseolus vulgaris*				

TABLE 3. Nitrogen Balance in the Biomass (Vegetative) Component in a Corn/Bean Polyculture Compared to a Corn Monoculture in the Tropical Lowlands of Southeastern Mexico

Source: Adapted from Stephen R. Gliessman, "Nitrogen Distribution in Several Traditional Agro-Ecosystems in the Humid Tropical Lowlands of South-Eastern Mexico," *Plant and Soil 67*, no. 1 (1982): 105–117.

done primarily through custom or superstition. But when examined in detail, some of these practices have a strong foundation in ecological principles. For example, a large beetle (*Epicauta* sp.) is well known in Tabasco for its ability to invade and quickly defoliate bean crops. When they appear, the beetles are gathered by hand during the early morning, when they are most active. Then they are placed in pots or jars and kept captive through the rest of the day. Late in the afternoon, the pots are heated over a flame just enough to kill the beetles, but not burn them. The pots, with the dead bodies, are then returned to the field and placed about twenty meters apart throughout the planting. The next morning, any live beetles left in the field are gone, apparently frightened away by some odor produced during the killing process.[10] Studies have shown that when some insects are attacked they release such a substance, known as an alarm pheromone, to alert others of their species to potential danger; research to develop these products for pest management is underway.[11] Unfortunately, few farmers use this method today in Tabasco because of the easy access to modern pesticides.

Weeds are also at a disadvantage in mixed cropping systems.[12] In a system where no herbicides are used, it is important to minimize the space between crop plants available for noncrop species—what we commonly call weeds. It was interesting to us that local farmers referred to squash more for its importance in weed control under corn than for any food it produced.[13] Further research showed us that a corn crop allows considerable light to penetrate to the soil surface, but that squash, with its large, thick, horizontally placed leaves, creates a dense shade that allows little light through. In addition, chemicals produced by the squash leaves are washed into the soil by rainfall, acting as a natural herbicide for weed species (a process known as allelopathy). Thus, through competition for light and chemical inhibition from the leaves, weeds are effectively controlled.[14]

In the long term, a polyculture cropping system can lead to more efficient use of resources and, at the same time, can better adapt to greater variability in environmental conditions.[15] With each crop having different requirements, a mixture of species ensures that the component parts will probably not come into conflict for the same resources. In fact, with several species planted together, variability in the field conditions will have less of an effect on the overall system yield. For example, with the three crops closely spaced in a corn, beans, and squash system, soil conditions at any given planting site are likely to meet the requirements for at least one of the three species, ensuring a harvestable yield of at least one of the crops from all parts of the system. Such a built-in adaptability to variability allows the farmer to work more within the limitations posed by the environment rather than strive for uniformity, usually at the cost of considerable energy in the short term and soil quality in the long term, by not having to disturb the soil as frequently or leave areas exposed to erosion and nutrient loss.

Ecologically, then, many traditional agroecosystem management strategies combine high species, structural and temporal diversity, efficient nutrient cycling and energy flow, and an intricate complexity of biological interactions. Such complexities have been selected over a long period of time in response to a wide array of cultural demands. These systems are essentially waiting to be "rediscovered" and analyzed. The persistence of these cropping systems is a reflection of their flexibility, an aspect of agroecosystem design and management that needs to be given greater emphasis as we conduct research toward sustainable agriculture.

What we have learned from our studies of traditional agroecosystem design and management urgently needs to be used to preserve and improve agriculture for traditional cultures; at the same time, our findings can teach us much about orienting research on alternatives for our own agriculture.

For example, many of the advantages of polycultures observed in our studies in tropical Mexico are now being corroborated by research in progress in California.[16] In a series of experiments involving row crop vegetables, especially cole crops (such as cabbage and kale) and legumes (such as peas and beans), we have found a reduction in plant-feeding insects when we interplant with crops that are not attractive to the pest or when we allow limited weed growth around the crop plants. This reduction may be due in part to the ability of the interplanted species to "hide" the crop plant from the pests by interfering with their pattern of perception. Greater vegetational diversity in a field could also encourage higher populations and more activity of natural enemies, or present site-specific variability of light, temperature, or humidity that would be unfavorable for the pest species.

When we planted collards and bush beans together and compared them to each planted separately, populations of the common collard pest, flea beetles, were much lower in the polyculture system. In addition, we applied a variety of weeding regimes to both monoculture and polyculture plots and discovered that the presence of weeds in the system also lowered pest populations (Table 4). Similar patterns were true for the common cabbage aphid as well.

We also found that the polyculture cropping pattern can actively suppress weed growth. In the same collard and bean polyculture, both the number and total biomass of the weeds were suppressed (Table 5), especially compared to the slower-growing collards planted alone. This suppression may result from the crops' use of available resources, such as nu-

Cropping System	Number of flea beetles		Damaged leaves per collard [c]
	per 10 collards [a]	per 5 weeds [b]	
Collard monoculture			
Weed free all season	34.0a	—	54.4a
Weedy all season	6.6b	25.0	29.9b
Collard and bean polyculture			
Weed free all season	2.3c	—	34.1b
Weedy all season	0.6c	15.0	32.1b

TABLE 4. Flea Beetles (*Phyllotreta cruciferae*) in Various Collard Cropping Systems in Santa Cruz, California

[a] Means followed by the same letter in each column are not significantly different ($p = 0.05$). (All means are averages of three sampling dates.)

[b] *Brassica* spp. weeds.

[c] Percentage of leaves in each collard plant with insect damage.

Source: Data from Gliessman and Altieri in *California Agriculture* (note 16).

trients, light, or water. Further research certainly needs to search for crop combinations that exhibit highly competitive characteristics against weeds.

To determine which weeds might be beneficial to leave in a cropping system, we have been running trials using different weed species common to agriculture in our region. We have found that many weeds produce toxic chemicals that are washed off the leaves, exuded from the roots, or retained after decomposition in the soil.[17] We have also found that different crops show differing degrees of susceptibility to the toxins. Considerable research is necessary before we are able to establish the basis for manipulating weed populations. Eventually, we will have enough information to consider a crop and its accompanying weed species as integral parts of the same system, thus establishing criteria for potential management strategies.

Therefore, one can imagine a complex farm system using crop and non-crop species, occupying all available sites within the agroecosystem, and changing through the season as conditions shift. Our understanding of the requirements of each part of the system would be complete enough that both competitive and allelopathic limitation of one species by another does not take place. Rather than setting up direct interference, we can imagine a sharing of resources, stimulation of one component by another, and the long-term integration of all the parts. In an ecological sense, the

| | Weed Population | |
Cropping System	Number of Plants	Dry Weight (grams)
Beans		
Weed free	0	0
4 weeks weed free	78	64.1
2 weeks weed free	127	205.5
Weedy	90	302.1
Collards		
Weed free	0	0
4 weeks weed free	42	52.3
2 weeks weed free	52	55.2
Weedy	142	438.2
Beans and collards		
Weed free	0	0
4 wccks weed free	28	25.7
2 weeks weed free	84	93.2
Weedy	85	234.5

TABLE 5. Weeds in Collard and Bean Test Plots, Santa Cruz, California

Note: Weeding treatments, in plots five meters square, were established by selective hoeing. Plots were weed free or weedy all season or were kept weed free for four or two weeks after crop emergence, then allowed to become weedy.

Source: Data from Gliessman and Altieri, "Advantages of Polyculture Cropping" (note 16).

niche of each species is defined not only by where an organism lives and what it needs to survive but also by how it might benefit the long-term productivity of the entire system.

Although they are not evident in the polyculture cropping sytems, several other characteristics observed in natural ecosystems offer guidance for research in agriculture. One ecological principle offering numerous opportunities for applications in agriculture is the island biogeography theory.[18] We can visualize the crop system as an island separated by an ocean barrier made up of other crop systems or even noncrop vegetation. The ability of pest organisms to cross that barrier will be determined by the ecological characteristics of the barrier and the limits and tolerances of the organism. We can even imagine the barrier being effective enough to function as a pest control mechanism. At Santa Cruz, California, research is in progress testing such weed borders, either for their ability to block the movement of insects or to attract beneficial insects capable of controlling the unwanted species.

Another characteristic of natural ecosystems that needs to be researched in an agricultural context is the idea of system succession or regeneration. In nature, when some disturbance like a flood or a fire happens, the system soon recovers and eventually returns to a state of development similar to what it was before the disturbance. In this sense, we can consider cultivation practices and harvest activities as types of disturbances. By understanding how an agricultural system recovers from disturbance, how cycles are closed and energy flow becomes more efficient, we may discover means of replacing lost materials without depending entirely on artificial inputs from outside the system. A large component of sustainability lies in the ability of a system to regenerate itself following human interference. Research needs to focus more on the role a variety of species plays in contributing directly to the maintenance of productivity, rather than strictly on their contribution to the crop material actually harvested and removed from the system.

Of course, an agroecological approach is more than just ecology applied to agriculture. Unlike a natural ecosystem, an agroecosystem is constantly affected by human intervention. Hence the interdisciplinary perspective of agroecology also encompasses the field of cultural ecology. The development of agroecosystems is then seen as a process of coevolution between culture and environment, where the two constantly interact and evolve, one affecting the other and both together selecting for the technologies that are applied to food-production systems. An interdependence between culture and environment develops where the productive potential of agroecosystems is kept within sustainable limits. But as agriculture has become increasingly viewed as purely a production system and linked more closely with economics, we have lost sight of the strong ecological foundation upon which agriculture originally developed.

In a restricted sense, an agroecological approach is the science of ecology applied to solving agricultural production problems. Agroecologists study the environmental background of the agroecosystem, as well as the complex of processes involved in the maintenance of long-term productivity. A broad goal of such an approach is to understand how cropping systems have evolved, how they operate, and where improvements can be made. This goal is in direct contrast to a restricted agronomic emphasis on individual components and a preoccupation with the harvestable end-product rather than a concern for how productivity is established. Both ecologists and agronomists are concerned with the component parts of the cropping systems they study, but agroecologists base their approach on greater awareness of how ecological relationships function within the context of the agroecosystem. An interdisciplinary approach is critical as we strive to gain an understanding of these relationships. Studies of traditional, rural cultures where empirical knowledge has been gained through a process of trial and observation have taught us much about the ecological component of agroecosystem design and management and how interdependent it is with the cultural components. Improvements upon these traditional systems, or the development of new or alternative systems for the future, will involve the integration of ecological and cultural knowledge. Only in this manner can agriculture establish a truly sustainable base.

14

Replenishing Desert Agriculture with Native Plants and Their Symbionts

Gary Paul Nabhan

Replenishing means making full or whole again. To replenish, we must first admit loss. To justify adding stock back into a herd, a fish pond, or a plant gene pool, we must recognize that some critical element is missing that is needed to sustain it in the long run. To find the motivation to fill up the larder, we must acknowledge that it has been depleted and that somehow we have become sorrier for it.

Modern agriculture has let (temporarily) cheap petrochemicals substitute for the natural intelligence—the stored genetic and ecological information—in self-adjusting biological communities.[1] In deserts, this sort of substitution has taken an additional, tragic twist within the past six to eight decades. Desert farmers used to depend upon a wide range of native plants suited to water-limited conditions by virtue of their drought and heat hardiness. But with the arrival of the deep-sucking mechanized pump, these stress-tolerant crop varieties were abandoned as focus was shifted to the few cash crops whose yields (until recently) appeared ever more responsive to additional irrigation water and the liquid fertilizers it carried. Fossil fuel was used to pump fossil water laid down during the Pleistocene to irrigate crops adapted to humid conditions; these crops seemingly made the deserts bloom. Why grow modest-yielding crops that could endure late season drought when you could pump enough water to keep stress from ever setting in?

The "ever" has not lasted very long. Today, even the descendants of the pumping pioneers recognize the irreplaceable resources lost in just two generations. A Mexican friend of mine is the grandson of the man who first bought a pump to use near the Sonoran Desert coast of the Sea of Cortez in 1906. Originally drawing upon wood-fueled steam power, but quickly switching to diesel fuel, the man pumped sweet water from forty meters below the desert floor. His success with irrigated fiber and feed crops encouraged the expansion of the Costa de Hermosillo, Sonora, irrigation district into the 118,000 hectares of groundwater-dependent agriculture that it is today. The grandson now pays the price for this pioneer's success: the same well does not reach water for more than 100 meters, and the water is of poorer quality. The irrigation district has seen the abandonment of farm lands closest to the Sea of Cortez because salt water has intruded into wells there. Of the 800 million cubic meters of groundwater extracted there in a recent year a deficit of 450 million cubic meters was not naturally recharged, and the rate at which the water level has been dropping has accelerated.

The greater the depth of the water table, the more energy is needed per meter of lift. Primarily because of irrigation-related energy costs, Arizona's total agricultural energy use per hectare has been 500 percent of the national average. In some desert areas, pumping costs alone increased tenfold in less than a decade. Growing humid-adapted crops takes 20–40 percent more water to produce the same yields in arid lands as in milder climates where evaporation is not as great. Because desert farmers' production costs for conventional crops are so much higher than those of farmers in other regions, many have now begun to consider growing specialty crops that require less irrigation and are better adapted to stress.[2]

Yet few of the water-conservative native crop varieties common in deserts at the turn of the century are available today. Many have already become extinct due to earlier neglect.[3] To replenish his fields with plants that need only minimal irrigation under arid conditions, my Mexican friend feels he has little choice but to newly domesticate wild desert plants. Three woody desert perennials with which he is experimenting—jojoba, Mexican oregano, and chiltepines—require less than half the irrigation water per year that a single season of his father's annual crops would use. However, as monocultural, furrow-irrigated plantings, these desert species are vulnerable to diseases that hardly affect populations in the wild.

Because of such problems, it is time to look again at the diverse cactus forest of the Sonoran Desert, rather than focus upon a few of its more valuable trees. Even in water-limited environments, natural communities are endowed with nutrient-pumping plants, biological control agents, pollinators, seed protectors, nitrogen-fixing and decomposing bacteria, and a variety of other organisms performing various functions that benefit the community as a whole. We must reopen the floodgates, not for more irrigation water but for an influx of desert-adapted organisms, which in association with one another and ourselves may correct the course of agriculture in hot, dry regions.

RETURNING TO THE WILDERNESS FOR ANSWERS

Agricultural ecologists George Cox and Michael Atkins suggest that the kinds of agriculture with the best chance to endure are those that deviate least from the energy flows and nutrient cycles of the natural communities within which they work.[4] For deserts, little energy can flow and few nutrients can cycle until water flows. Israeli ecologist Noy-Meir observed that solar energy use by plants is so tightly coupled with rainfall and floods in deserts that "the vegetation in an arid system may be regarded as a converter of a water inflow to an energy inflow."[5] For this reason, agroecological studies in deserts will continue to emphasize plant and water relations, just as tropical agroecology is preoccupied with plant and insect relations.[6]

The variability in rainfall between seasons shapes wild desert communities as dramatically as does its paucity. Native plants remain dormant or minimally active during the seasons of more predictable drought. They are physiologically triggered to be responsive during typical periods when rain exceeds evaporation via temperature and photoperiodic controls of germination, leaf expansion, flowering, and fruiting. Conventional agriculture attempts to continue production during drier periods of the year by utilizing day-neutral plants that have less specific physiological triggers. Yet crop production should be scheduled with the rainy seasons not only because of a more favorable water balance. If crops continue to be irrigated in fields during periods when the surrounding desert is dry, they suffer from both climatological and biological "oasis effects": when a hot wind blowing across dry vegetation reaches a green patch of crops, it aggravates transpiration losses from that oasislike patch. The crops also become a sink for both invertebrate and vertebrate "pests" that will migrate in from the surrounding desert to feed upon the crops.[7]

Water is not equally distributed among desert habitats. During infrequent storms, much of the land area sheds a large percentage of the rain hitting it, and this runs off to floodplains where the water becomes naturally concentrated. Both floodwaters and the humus they carry are eventually deposited upon alluvial fans, which are unusually endowed with soil moisture and fertility. Certain alluvial fans have been demonstrated to be twenty-five times as productive as less favorable sites in wet years and even five times as productive as these other sites in years of severe drought.[8] Despite the fact that these watercourses may flow only a few days each year, they are key to desert productivity.

The Papago and other native cultures of the desert sought these alluvial fans for their fields. In one historic Papago community, 100 families maintained 355 hectares of crops on fans receiving storm water, organic matter, and nutrients from 240 square kilometers of watershed. There are perhaps 40,000 to 50,000 hectares of alluvial-fan habitat suitable for Papago-style floodwater farming in the northern third of the Sonoran Desert, or the area of southern Arizona below the Mogollon Rim. If these runoff-fed sites were brought into cultivation, they could potentially produce food on about the same amount of land reportedly taken out of conventionally irrigated agriculture in Arizona between 1975 and 1980.[9]

It is worthwhile to place agriculture within these desert floodplain habitats, where organic matter and nutrients have naturally accumulated for millenia, and where, if watersheds are properly managed, they will continue to accumulate. In the Sonoran, Mojave, and Chihuahuan deserts, the uplands are often dominated by woody legumes such as mesquite, palo verde, ironwood, and acacia, several of which fix nitrogen. With a single, intense storm, enough nitrogen-rich litter from these trees, rodent feces, and other decomposed detritus is shed onto the alluvial fans to add as much as thirty cubic meters of organic material to each hectare. This desert organic debris ranges in total nitrogen content from .79 to 1.13 percent (livestock manures, by comparison, contain .25 to .60 percent nitrogen).[10]

Whereas organic matter in North American desert soils normally ranges from .25 to .69 percent, alluvial soils on the Papago Indian Reservation that receive flood-washed debris average 1.25 percent organic matter.[11] Despite centuries of cultivation without the use of livestock manures or chemical fertilizers, the Papago field soils are not significantly different in nitrogen or organic matter content from noncultivated alluvial-fan soils nearby. At certain sites, the Papago encourage the deposition and accumulation of this organic matter rather than depleting it, to the degree that the soils compare well with semiarid grassland soils (Figure 1).[12]

However, this natural renewal of moisture-holding organic matter and associated nutrients is easily disrupted by overgrazing and by stream-course

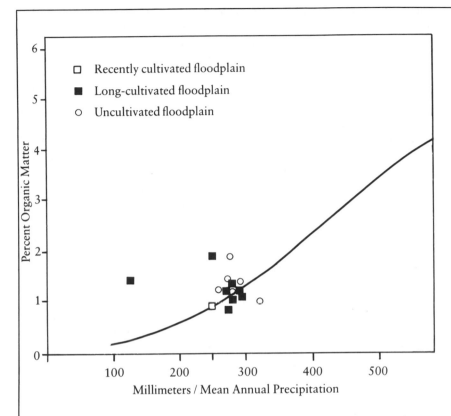

FIGURE 1: Estimated annual rainfall and organic matter percentages for Sonoran Desert alluvial fans in or near Papago Indian fields compared with a standard relationship for the semiarid West.

Source: Adapted from description in Thompson and Troeh (note 12).

manipulations that increase the "flashiness," or rapidity of floods. Alluvial-fan agriculture is dependent upon healthy watersheds. Consequently, upstream rangelands not suitable for cultivation must be managed so that grazing does not outstrip the growth of vegetation there and depress potential yields in the few favorable years the desert has. Using these ecological guidelines for managing agricultural fields in well-selected sites, we have a better chance of sustaining productivity than if we try to make the whole desert bloom with industrial techniques and products.

THE SONORAN DESERT: PLANT COMMUNITIES OF DIVERSE LIFE FORMS

Walking through a saguaro cactus "forest" in the Sonoran Desert, you quickly realize that the cacti do not dominate this kind of vegetation to the same extent that perennial grasses cover the prairies or ponderosa pines control the sierran montane conifer forests. Saguaro and prickly pear cacti seldom cover much of the ground, but their shallow lateral roots (four to ten centimeters deep), which rapidly absorb even light rainfall, extend up to three meters from the bases of their stems. These succulents may be intermixed with tree legumes such as mesquite and palo verde that have incredibly deep taproots capable of pumping up both subsurface water and nutrients. Small shrubs such as brittlebush have more generalized root systems, with taproots one-half meter deep and branching lateral roots running horizontally fifteen to thirty centimeters below the soil surface. This puts them above palo verde roots but below prickly pear roots.[13]

In summer, the shallow-rooted perennial plants compete for available water and nutrients with fast-growing annuals that can complete their life cycles in six to nine weeks, before the drought sets in. Many of these plants—such as amaranths and six-weeks grama grass utilize the water-efficient C_4 photosynthetic pathway that works well under high light and temperatures. Vining herbs, such as wild gourds, emerge from perennial taproots during the summer as well. In the winter, when these herbs are dormant, an additional set of annuals, such as chia and tansy mustard, hug the desert floor.

Ecologist Otto Solbrig has suggested that "this variety of life forms indicates that there are several competitively viable solutions to the problems of survival and reproduction in deserts."[14] In some years, drought-escaping annuals outproduce all other life forms.[15] In other years, few annuals even germinate, and the deep-rooted trees or such water-conservative succulents as cacti and agaves are responsible for the bulk of the production.

Mixtures of species of life forms have been shown to have more stable yields than homogeneous stands of one genotype because they partition environmental resources more efficiently in time and space. But ecologist R. W. Snaydon has argued that the benefits of diverse mixtures are even greater under varying conditions where suboptimal conditions put otherwise high-yielding genotypes under stress.[16] In hot climates where rainfall is unpredictable, it is no wonder that natural communities are a balance of life forms with extremely different strategies for coping with stress.

Fortunately, botanist Richard Felger has discovered promising orchard

Life-Form Classification and Species	Attributes/Economic Products	Virtues in Association with Domesticates
Herbaceous Perennial		
Coyote gourd, *Cucurbita digitata*	C₃ drought and heat tolerant; sizable with low consumptive water use; seeds rich in protein and polyunsaturated oil; tolerant of vine borer; also medicinal.	Pollinated by and encourages presence of native bees that also efficiently pollinate cushaw squash; may also serve as a vine borer "sink" to reduce predation on squash.
Metcalfe's bean, *Phaseolus metcalfei* and *P. ritensis*	C₃ drought tolerant, quick-growing vines; large, protein-rich beans; good forage; root is a traditional medicine; nitrogen fixing with native *Rhizobia*.	May encourage persistence of native *Rhizobia* that will also serve annual (tepary, lima) beans; also cross-compatible with limas; erosion control.
Desert unicorn plant, *Proboscidea altheaefolia*	C₃ drought and heat tolerant; root is a traditional medicine; seeds rich in protein and polyunsaturated oil.	Pollinated by and encourages presence of native bees that also efficiently pollinate annual devil's claw crop.
Cañaigre, *Rumex hymenosepalus*	C₃ flower stalks like rhubarb, and seed protein rich for flour; root formerly an industrial source of tannin and medicine.	Indicator of low soil potassium; encourages presence of parasitic Hymenoptera that control pest populations.
Xerophytic Trees and Shrubs		
Chiltepine, *Capsicum annuum* var. *minimum*	C₃ delicious ball-like hot chiles already a cash crop in northern Mexico; heat and virus tolerant.	Cross-compatible with larger chiles; proven source of virus resistance.
Mesquite, *Prosopis velutina*	C₃ pods a major, caroblike wild harvest historically; protein-rich seed; excellent wood for crafts and fuel; large crops 4 out of 5 years.	A nitrogen-fixer and -pumper producing litter in abundance that can serve as compost or mulch for annuals; excellent hedgerow, windbreak, and wildlife food.
Wolfberry, *Lycium* spp.	C₃ tart fruit harvested in quantity for jelly and syrups by Pima today; drought and heat tolerant.	Indicator of good soil for floodwater annual crop production; excellent hedgerow plant and food for wildlife.
Cacti		
Prickly pear, *Opuntia phaecantha* or *O. ficus-indica*	CAM* drought and heat resistant; sweet juicy fruit already an economic crop for "cactus jelly."	Vegetatively propagated easily as living fence between annuals.
Organ pipe, *Stenocereus thurberi*	CAM* drought and heat resistant; fruit marketed fresh in Mexico, also suitable	Living fence plant on field edges; roost for insectivorous birds.

Bacanora mescal, *Agave angustifolia* var. *pacifica*	CAM* drought and heat tolerant; rapid water uptake after rains; caudex can be roasted and processed into many products, including tequilalike beverage.	Excellent erosion control plant in rows at edge of terraces; cross-compatible with tequila agave.
Palm		
California fan palm, *Washingtonia filifera*	CAM* heat resistant; fruit, trunk wood, and fronds major resources historically for a variety of products.	Pollen utilized for fertilizing date fruit; can create an oasis microclimate for other plants and insectivorous birds.
Summer Ephemerals		
Sonoran panic grass, *Panicum sonorum*	C_4 milletlike cereal suitable for tamales, pinole, and crepes; water-use efficient, rapid yielder adapted to floodwaters.	Quick-growing cereal to be intercropped between perennial strips.
Desert amaranths, *Amaranthus palmeri* and *A. cruentus*	C_4 leafy vegetables with high lysine seed; water-use efficient, rapid yielder adapted to floodwaters.	Harbors alternative prey that help maintain population sizes of pest predators; indicator of soil nitrogen levels.
Devil's claw, *Proboscidea parviflora* var. *hohokamiana*	C_3 heat tolerant, rapid yielder of seeds rich in protein and oil; fruit fiber a cash crop in Indian basket industry.	Quick-growing seed crop to be intercropped between perennial strips.
Tepary bean, *Phaseolus acutifolius*	C_3 heat and drought tolerant; protein-rich edible bean and forage produced in abundance over a short season.	Quick-growing, nitrogen-fixing legume to be intercropped between perennial strips; reduces outbreaks of 3 diseases and 2 insects.
Winter Ephemerals		
Chia, *Salvia columbiarae*	C_3 drought tolerant; rapid producer of gelatinous, nutrient-rich seed.	Quick-growing seed crop to be intercropped between perennial strips.
Huazontle, *Chenopodium berlandieri*	C_3 drought tolerant; producer of mineral-rich, edible greens and flower head like miniature cauliflower.	Quick-growing vegetable to be intercropped between perennial strips; increases activities of parasitic Hymenoptera that control pests.

TABLE 1. Native Sonoran Plants to Replenish Agricultural Ecosystems
* Crassulean Acid Metabolism, the most water-frugal photosynthetic pathway.

or field-crop candidates in every set of Sonoran Desert life forms.[17] Many of the species with the most potential as nutritious food crops (Table 1) were major subsistence resources of ancient hunter-gatherers, and some were later used as living hedgerows by desert farmers.

For such native species, yield parameters for edible pods, high-value wood, honey, and pollen have already been investigated. With little or no input of water or fertilizer, a mature orchard of 118 mesquite trees per hectare can be expected to produce 2,300 kilograms of pods per year, as well as highly valued fuel wood. An eight-to-ten-year-old planting of 5,000 agave plants per hectare, depending upon the species, can produce 25,000 liters of tequilalike mescal, 22 tons of sisal fiber, or 125 tons of sweet, edible leaf bases and meristem tissue. The water-use efficiency of agaves is at least twice that of corn. Drought-escaping devil's claw plants, under monthly irrigation, can produce 2,330 kilograms per hectare of seed, for processing into 1,000 kilograms of highly polyunsaturated salad oil and into a seedmeal flour amounting to 675 kilograms of protein. These already economic yields, estimated for pure stands of desert crops, are likely to be even greater under polycultures where different life forms are intermixed in adjacent strips or mosaics so that pest and disease damage is reduced.[18]

Another way to replenish the natural intelligence of arid land agriculture is to focus cropping systems on genera of plants that are adapted to desert conditions and have in association with them a diversity of coevolved symbionts. In short, healthy desert agriculture is native not merely when native peoples are the farmers. It is native when crops are largely evolved from wild relatives that can still be found as volunteers in or near the fields. Because of their long evolution in the region, they attract highly adapted pollinators and root-associated microorganisms such as nitrogen-fixing bacteria and water-absorbing mycorrhizae. Long-associated pests may be present, but pest predators that have coevolved with them are likely to be found on native weeds (see Table 1). By encouraging the perennial wild relatives to persist in or near fields, we may allow for the continued flow of "hardy" genes into the crop species. Wild perennials may reinforce the presence of these symbionts, further aiding the crops. A few examples may help clarify the potential value of these relationships.

Tepary beans are a native desert crop that the Papago and other Indians have utilized for centuries, taking advantage of the resistance these plants have to drought, heat, salt, boron, two insects, and three diseases. Nodulation of their roots by nitrogen-fixing bacteria is unfortunately uncommon in conventionally irrigated fields. Even when the bean's roots are artificially inoculated with native strains of beneficial bacteria, these *Rhizobia* do not necessarily set in the extreme conditions of desert summers. However, nodules can be regularly found on teparies as well as their wild relatives growing near Papago fields. No one knows whether or not native *Rhizobia* strains wash into the fields from wild bean stands upstream, but we do know that the nodulated tepary plants produce far more abundantly than pinto beans without nodules that have been introduced in the same fields.[19]

Two other desert-adapted cultivars, striped cushaw squash (*Cucurbita mixta*) and devil's claw (*Proboscidea parviflora* var. *hohokamiana*), may benefit from wild relatives that maintain populations of solitary bees in and around fields. Wild perennial gourds (*Cucurbita digitata* and *C. foetissima*) have coevolved with two genera of bees, *Xenoglossa* and *Peponapis*, which are ideal farm workers.[20] They show great fidelity to cucurbit blossoms, including those of cushaw squash; the male bees sleep inside them and seldom visit other kinds of flowers as honey bees do. Dr. Steve Buchmann has found that they wake up earlier and move more rapidly and efficiently between the cucurbit flowers they are pollinating than do honey bees.

Similarly, a bee named *Perdita hurdi* has coevolved with a wild perennial devil's claw, *Proboscidea altheaefolia*. This bee cuts into the flower tube of unopened devil's claw blossoms just as the pollen becomes ripe and transfers it to other devil's claw flowers, ensuring cross-pollination.[21] I have also found these bees working domesticated devil's claw flowers in a Papago garden. Many native solitary bees have been driven to extinction by the invasion of the Old World honey bee, but the survivors are potentially much more efficient pollinators of certain plants by virtue of their fidelity to particular genera.

Making use of relationships such as these, native desert agriculture draws not only from ten millenia of plant domestication and selection by native Americans, but from the millions of years of natural selection and adaptation to desert environments of both plants and animals.

DESERTS: WHERE ADDICTS GO TO DRY OUT

The limitations of this ecological approach to agriculture may not be obvious to the optimistic reader. First, there is currently no place where one can see all these desert-adapted strategies integrated in a single field, although one or several of them are used in certain Papago and Sonoran mestizo fields. It is doubtful that any single desert field should use all these strategies, for no two fields are alike. The broad generalizations given here will not guide anyone in making any particular desert field work. One must simply let each distinctive field "speak" for itself; learning its language is more critical to good farming than reading any book.

Second, the kind of native desert farming discussed here cannot be done by just one farmer. Unless there is a community of farmers sharing observations, tools, and hope, it is bound to be unattainable. Just as the wild species must complement and reinforce one another for these field ecosystems to work, farmers' efforts must reinforce one another through time. Moreover, for watershed-based floodwater farming to survive, downstream cultivators must be in accord with upstream sheepherders and grazers. Such balances cannot be legislated. They must arise out of shared cultural ethics and experiences. Such responsibility to other desert dwellers will not be easy at a time when rural communities are disintegrating in Arizona and Sonora as elsewhere. I have made the remote, riverless desert environments the subject of this chapter not because I believe the Central Arizona Project and other irrigated agricultural districts are unimportant. The lands affected by these federally subsidized projects need committed, careful management as badly as other farmlands. Instead, I look to the riverless deserts because these marginal environments may have to absorb future agricultural expansion and because, as extreme environments, they clarify for us all the inescapable dependence of agriculture on its natural sources.

15 | Investigations into Perennial Polyculture

Wes Jackson
Marty Bender

We believe that the best agriculture for any region is the one that best mimics the region's natural ecosystems. Such an agriculture holds soil erosion down to the level where nature can replace what wind and water carry away. It copes successfully with insects and disease and it utilizes water in the optimum way. In our Great Plains region, the natural vegetation is prairie, mixed communities of grasses, legumes, and other plant families that within the past hundred years were supporting phenomenal amounts of animal and human life. The prairie is mostly gone now, victim of the plow and the international prices of corn and wheat. But our goal is to revive a version of it, to create prairielike grain fields, combinations we call herbaceous perennial seed-producing polycultures.

To that end, we at The Land Institute, situated in the wheat-farming country of central Kansas, have posed three biological questions. They guide our research now and we expect them to direct our research in the future. If we can find affirmative answers to the first two questions, they can have tremendous implications for the future of grain farming because such answers would fundamentally alter the agriculturist's understanding of the world of domesticated plants. The third question follows from the first two. These questions are:

• Can herbaceous perennialism and high yearly seed yield go together?

• Can herbaceous perennial polycultures have an economic advantage over herbaceous perennial monocultures? (We know that annual polycultures can outyield annual monocultures.)

• What are the major considerations for managing domestic prairies and marshes, that is, crops composed of herbaceous perennial polycultures?

We have surveyed hundreds of species of perennial grasses. We have researched the literature, conducted experiments in the field, and grown the plants we judged most likely to give high yields of seeds. During this phase of our investigation, begun in 1978, we have paid little attention to the utility of the plants we have studied; we have not been concerned with whether they could be directly useful to human enterprises. At stake, instead, was the simple, principle question of whether high yield is possible in perennial flowering plants.

The first question goes right to the heart of a fundamental consideration in biology. Many biologists assume that a plant is effectively a closed energy system of roots, stems, leaves, and reproductive material (the flowers that result in fruits and seeds). Stems and leaves are energy poor, in most cases; plants pour most of their energy into the root and the seed.

If the roots have most of the energy, the plant tends to be perennial and polycarpic. In other words, it can overwinter and set seed a number of times in its life. Polycarpic plants generally don't allocate much energy to seeds. Our major food crops—wheat, corn, rice, and oats—are monocarpic plants.

Monocarps tend to send lots of energy to seeds—like wheat grains and ears of corn—once in their lives and then die. Because these plants are annuals, the farmer must go out and replant them each year. We want to see whether a farmer can replant polycarpic plants just once every three, or five, or ten years and take out a good harvest every year.

Some biologists are sympathetic with our work, but do not necessarily believe we will succeed. They believe that as we move toward high yield in our breeding program, we will automatically be moving toward annualism. Their reasoning—the received wisdom of a whole tradition of plant science—is that a plant has only so much energy and that if we want to breed for more energy in the seeds, we will have to "steal" the energy from the roots. As we increase yield, we will be moving toward energy-starved roots, and the plant won't be able to survive the winter. In effect, we would be converting a perennial plant into an annual.

We are questioning this "closed system" model and are optimistic for two reasons. One reason comes from history. Corn averaged about 30 bushels per acre in 1930 and now averages around 100 bushels per acre. One-half of this 70-bushel increase comes from changes in the farming environment (introduction of commercial fertilizer, improved methods, and so on), according to Dr. William Brown, recently retired president of the Pioneer Seed Company of Johnson, Iowa. The other 35 bushels are attributable to genetic improvement—but how, precisely? The corn plant is about the same height and the leaves are nearly the same as they were in 1930. The 117 percent improvement did not come from converting a perennial into an annual, for corn was an annual to begin with. There was no extra root energy to steal. All that extra yield had to come from somewhere else.

The more important reason for optimism comes from our work with the Maximillian sunflower. In the spring of 1981, we planted two rows of the species, fifty feet long, three feet apart. These plants cast seed at season's end, in the fall of 1981. In the spring of 1982, we noticed seedlings emerging in tilled land immediately south of their parents. We thought it would be good to compare these two populations, so we defined two more rows with a rototiller. Remember that the plants did not have to wait on us to be planted; the seeds had been naturally thrown in the fall and they germinated and grew as fast as nature would allow. In other words, they got the earliest possible start in the spring of 1982. Yet at no time during the 1982 growing season did these 1982 plants catch up with their parents, which had been planted in 1981. The one-year-old plants had stored an abundance of sunlight in their roots and used some of it for fast early growth, and they kept the lead.

The significance of this energy storage has to do with energy harvest. The parent plants, which had covered their whole plot, were able to capture more sunlight than their offspring could, even though the offspring had germinated as early as possible. If there is more total energy harvest by a plant that gets started early in the spring, then it becomes a question of how that extra energy is allocated within the plant. If more is allocated to roots, stems, and leaves than is necessary for their livelihood, it seems possible to select plants that allocate more to the seed. In other words, if the "extra" energy captured can be sent to the seed, then the plant should have enough to support the other plant parts and additionally give us higher grain yield. Because more energy is harvested by a plant that gets started early in the spring, it seems to us that, through breeding, we can develop plants that will allocate this extra resource to the seed without "robbing" the root.

This example is a minor one compared to the real challenge we wish to make, but we think that even with the closed system model, some of the resource ordinarily allocated to roots or rhizomes could be sent to seeds. We doubt that, in most cases, the energy stored in roots is so marginal that a perennial would be in danger of becoming annual if a little energy were "stolen." So, in short, we believe the potential is there. The next question is, how great is that potential?

Historically, there has been little incentive to develop high seed yields in perennials. For millenia, the big effort has gone into developing high-yield annuals. Companies and individuals who sell native plant seeds have always had more reason than most to increase yield in their perennials, but to them and their customers, even a modest increase in germination has been more important than an increase in yield. The Soil Conservation Service's regional plant materials centers are woefully understaffed and have little time to devote to increasing yield. Range agronomists have understandably paid close attention to forage yield increases and have put only minor efforts into seed-yield improvement. The universities have had other agendas. In short, there has never been anything close to an aggressive program for increasing herbaceous perennial seed yields.

As we began our exploration at The Land Institute, we started by defining what we meant by high yields, using winter wheat as our standard. It will average around 1,800 pounds per acre (30 bushels at 60 pounds per bushel). Then we set out to learn what the highest recorded yields of herbaceous perennials were. Our survey of the literature on herbaceous perennial yields revealed the following: Buffalo grass (*Buchloë dactyloides*) that had been fertilized and irrigated yielded 1,727 pounds per acre. This figure includes burs, of which seeds are a small part, but the yield does represent fruit and seed material, just as the wheat grain does. Fertilized and irrigated alta fescue (*Festuca arundinacea*) averaged 1,460 pounds per acre. A native stand of sand dropseed (*Sporobolus cryptandrus*), under dryland conditions at Hays, Kansas, yielded 900 pounds per acre.[1]

Irrigated legume yields were also encouraging. A five-year average for Illinois bundleflower (*Desmanthus illionensis*) amounted to 1,189 pounds per acre. Fertilized cicer milk vetch (*Astragalus cicer*) yielded 1,000 pounds per acre, as did sanfoin (*Onobrychis viviaefolia*), which received seventy pounds of nitrogen per acre.[2]

These high yields come from plant materials that have had little or no selection for high seed yield. These are species in which yield may be somewhat less important than the quantity of protein per acre. The Illinois bundleflower, cicer milk vetch, and sanfoin samples all have a protein yield of about 400 pounds per acre.[3] This exceeds the protein (228 pounds) from an acre of wheat yielding 31.6 bushels. It even compares favorably with 100-bushel-per-acre corn, which would yield about 500 pounds of protein per acre.

At The Land Institute, we have various species planted for the specific purpose of obtaining yield data over a period of several years. Because plants growing on an edge or near a margin tend to be more vigorous than those in the middle of a stand, it is not entirely legitimate to extrapolate from a row five meters long and one meter wide. If it were legitimate, though, then an acre of wild senna (*Cassia marilandica*) could be expected to yield over 4,200 pounds of seeds in its third year. We mention this to illustrate that even third-year plants, at least in such an herbary row, are still high yielders. Seed from first-year plants that had been planted late but harvested from an eight-meter row in the midst of a field of wild senna averaged 2,300 pounds per acre. Curly dock (*Rumex crispus*) growing wild yielded over 4,400 pounds of seed per acre.

As more data on perennial seed yields are collected, we expect to discover numerous other perennials with high-yield potential. For example, almost all perennial seed yields mentioned here are from grasses ordinarily used for forage and hay. The breeding that has been done has sent energy to the leaves rather than to the seed. Perhaps there are perennial grasses, poor for forage but good for seed yield, that have not yet been studied. The dropseed genus *Sporobolus* and the perennial ryes of the genus *Elymus* may include such grasses.

The direction of future research on perennial grain crops may depend upon the results of research yet to be done. Such research could determine which perennials grow better in rows and which ones should be sown or broadcast. Our first motivation for researching perennial grain crops was to prevent soil erosion beyond natural replacement levels. Consequently, we need detailed studies to determine the relative ability of various perennial grasses and broad-leaved plants in rows to control erosion compared with such plants in solid stands. Perennial grasses grown at the institute on a hillside in rows three feet apart with clean cultivation in between have substantially reduced soil erosion. It is possible that rows of certain perennials will not reduce soil erosion to a sustainable level, so seed yields of perennials in solid stands will need to be recorded.

The Land Institute's selection program is minuscule, and it does not compare to the work being done on annual crops. But the improvement in perennial yields is sure to be greater than can be expected in future improvement of annuals. The gains will come from most of the same sources breeders historically have called upon to increase yield: the reallocation of resources within the plant and the increase of plant performance overall.

To be a good candidate for agricultural adoption, a perennial species should: (1) demonstrate high variation so that the breeder has a basis for selection; (2) show potential as a high seed yielder (it might also be a strong nitrogen fixer, for instance, or a natural repeller of pathogens and pests); (3) have seeds that mature more or less simultaneously, are shatter resistant, and have a favorable ratio of reproductive shoots to vegetative shoots; (4) promise a relatively high yield for a minimum of three years; and (5) exhibit stability in chromosome assortment so that seed fertility can be high.

COMPETITION: MONOCARPIC AND POLYCARPIC PLANTS

At one level, the central question of our work is whether herbaceous perennials can give enough yield over the years to compete with annual grain crops. Perhaps a more fundamental or scientifically correct way to ask the question is, "Can a polycarpic plant compete in yield with a monocarpic plant?" A couple of questions come to mind when we think of a plant as polycarpic rather than as perennial:

1. Are the monocarpic and polycarpic characteristics fundamentally opposed, or do they occur within a genetic continuum that might be manipulated, perhaps with ease?

2. If it is not a fundamental opposition (and we are guessing that it isn't), are these characteristics controlled by the same genetic assemblies in all herbaceous flowering plants, or are there numerous genetic routes to the same end? If they are not controlled by common genes, are they common at the family level, the generic level, or even the species level?

The answers to these questions will require years of research, but they are of great importance, especially if we want to convert our domestic crops into perennials.

An herbaceous polycarp will reserve energy in its roots or rhizomes for the next year's growth. An annual herbaceous monocarp won't. A perennial herbaceous monocarp, such as agave, the century plant of the Southwest and Mexico, has to carry energy over from one year until the next. When it finally blooms after twenty-five years or so, it dies. Natural selection likely favors the two extremes of annual monocarpism and polycarpism. It would be wasteful for a monocarp to retain much energy or many nutrients in living vegetative matter after it sets seed. The strategy of the polycarp, on the other hand, would be to conserve energy and nutrients.

MANAGING THE DOMESTIC ECOSYSTEM

In the long run, breeding high-yielding polycarpic perennials may be the least difficult problem in developing a sustainable agriculture. Getting a perennial-based agriculture operating would be complex, too. There is the problem of achieving the right species mix for a particular region or farm or field. Different farmers have different preferences and needs. One piece of ground is not like another. The problems go on and on. The crucial point is that the problems of monoculture have grown so great that they make it desirable to tackle the problems of polyculture.

In a review of the literature on annuals in polyculture, D.C.L. Kass concluded that polyculture is beneficial, especially if appropriate crops are chosen.[4] Getting the right mix appears to be of critical importance. More specifically, Kass concluded that polycultures have clear benefits when we consider nutrient withdrawal from the soil. When one of the crops is a legume, the nitrogen in the soil-plant system increases. All this leads, Kass contends, to greater stability in yields as time goes on. This should be true of perennial polycultures as well.

Managing and husbanding a perennial polyculture may at first seem unacceptable to farmers. Such an agriculture may appear to be so complex that experts would be necessary, and no farmer wants to depend on experts for his or her livelihood. But management problems would be fewer than one might expect, as range management suggests. A well-managed range or pasture for forage production is sustainable, and it is our only domestic analog to the wild prairie. Good range management may require more thought (although we have some reservations as to whether it, in fact, does), but on a per-acre basis, it requires less energy and time than a corn or wheat field.

Jack Harlan commented that "the general principles that apply to the dynamics of natural grasslands apply just as well to the dynamic balance of species in tame pastures."[5] Likewise, management of perennial polycultures of grain crops would apply many of these same principles. The main difference would be in kinds of harvest methods.

Some grain-crop breeders warn that pest problems will mount in an herbaceous perennial polyculture, unless we use lots of pesticides. We know that without chemicals, pest problems can increase significantly when monocultures are planted in the same fields year in and year out. It is also true that insects and pathogens could build up, and weeds could creep into a polyculture mix. On the other hand, Walter Pickett, plant breeder at The Land Institute, has talked with grass breeders about this (most recently at national meetings held at Pennsylvania State University in 1982) and reports that we should not be concerned about the difficulty of pests in perennial polycultures.

The differences between the breeding strategy of traditional grain-crop breeders and that of forage breeders explain these opposite responses. It is perhaps a carry-over from our puritanical past that causes a wheat or corn breeder to go with one strong gene for resistance to a particular pest. This is easy to handle in a breeding program, but it is also easy for the pest to overcome. Consequently, breeders are forced to release one new strain after another, trying to stay ahead. It is the exclusion of genes in a mono-culture that helps breed diseases and pests. Forage breeders, on the other hand, breed for general resistance to a disease. They want to keep the pest the way it is. By keeping the crop genetically broad based, they keep the pest genetically adapted to everything. Thus, the pest is likely to be not well adapted to any particular thing. There are numerous twenty-year-old pastures around, and grass breeders and pasture experts seem to have few pest problems, even with minimal management.

We know that one-trait breeding or few-traits breeding in livestock al-ways brings trouble. Too exclusive a focus on production will sometimes bring problems in reproduction. Too exclusive a focus on speed or size or color in racing or show horses entails losses of qualities equally desirable or more so in the long run. The differences of kinds of desired perfor-mances are probably critical—for example, the differences between a good work horse and a good show horse. The good work horse must fit into a complex form; it must do many, varied things well. The good show horse must have good form, but the show ring is a simple and simplifying place. This is analogous to the difference between a desirable species in a peren-nial polyculture grain field and a modern wheat variety. The need is to de-sign these polycultures according to place or situation. Modern agricultur-ists design for an environment homogenized by commercial fertilizers, pesticides, irrigation, and even by certain tilling methods.

We don't think it is too bold to say that, in the long run, if a domestic ecosystem is properly managed, we can forget about "pest management." In the meantime, numerous strategies may have to be employed. It may even be desirable to keep highly susceptible plants in the mix as hosts for insects or pathogens, so that when a new strain of the pest arises in the larger population, it will be "watered down" quickly through genetic re-combination. Another way to look at it is that the pest keeps the crop pop-ulation genetically "toned up." Even though generalized resistance may lead to reduced yields, it is a sunlight-sponsored resistance not requiring fossil-fuel feedstocks or other expensive energy inputs, and the resistance is a chemically safe one besides.

This all makes sense if we imagine an agriculture sponsored totally by sunlight where all the cycles of nature are closed. Imagine, for example, a farm in the future that supports itself by the energy harvested from its own fields. It is a safe guess that if a domestic ecosystem sponsors generalized resistance to pests by relying on biological information (for example, genes that are responsible for immunity), the energy cost will be a small fraction of the cost if we control the pests with plowing and spraying. Remember that we are talking about a farm that is much more self-supporting than a modern industrial farm. If the traction energy for working the ground comes from the farm and goes through a draft animal, the energy cost will be high; if it comes from the same farm and goes through a tractor as an alcohol fuel, the energy cost will be higher still.[6] Likewise, the energy costs for plant protection seem certain to be lowest when the protection comes from within the roots, stems, leaves, flowers, and seeds, where we can take advantage of the efficiencies of miniaturization inherent within the plant.

BENEFITS OF AN ECOLOGICAL AGRICULTURE

At The Land Institute, we have envisioned eight benefits from an agriculture based on perennial plants in polyculture. Six of the benefits involve resource considerations; two are social.

1. We expect perennial agriculture first to cut net soil loss to zero and then to build soil. The fossil energy saving for fertilizer to replace this lost soil would be significant.

2. The agricultural consumption of fossil energy would substantially decline, even where there is no soil loss.

3. The perennial ecosystem would conserve water and use it at near maximum efficiency. Springs, long since dry or short lived, would return. The need for irrigation would decline, again reducing fossil energy use, both indirectly and on the farm.

4. Industrial pest-control chemicals, especially those with no close chemical relatives in nature, would no longer be necessary. This means far less chemical contamination and minor fossil energy savings.

5. The direct consumption of fossil energy in the field would be greatly reduced because annual seed bed preparation, annual planting, and cultivation for weed control would no longer be required.

6. Because of increased efficiency in water usage, irrigation problems—aquifer mining, water diversion, soil salting, siltation—would become more manageable.

7. Land ownership questions would become more tractable because of reduced expenses for machinery, energy, farm chemicals, irrigation, and seed.

8. Over 100 million acres once cultivated but now set aside as marginal

land—largely because they are vulnerable to erosion under current cropping methods—could be brought into production, thus reducing land
prices. This increases the opportunity for more people to have a farm as a
place to earn a decent living and to live decently.

Our original motivation for working toward an ecosystem agriculture
featuring mixtures of perennials was to save our soils from erosion or
chemical contamination.

Even if we do have to plow in order to replant, say every three years or
so, our perennial fields will lose less soil than a field that must be plowed
every year. The fibers of a grass sod remain tough for some time after
plowing and continue to help to bind the soil.

We have made a rough estimate of the energy in fertilizer required to
replace the nitrogen (N), phosphorus (P), and potassium (K) that are lost
to erosion over the 316 million acres devoted to the top ten crops in the
United States today. If soil loss averages only five tons per acre, the fertilizer has an energy value of about ninety-six million wellhead barrels.[7] This
is about equal to the energy value of the N, P, and K tied up in fifty-bushels-
per-acre corn over the same area. Assuming a five-year replant cycle and a
soil loss of five tons per acre during the year of plowing and no soil loss
during the other years, we could save the equivalent of eighty million wellhead barrels per year in the N, P, and K saved with reduced soil erosion.
Unfortunately, we cannot credit this energy savings because the energy
value of nutrients lost in soil erosion is not accounted for in current energy
analyses of American agriculture.

Although the primary purpose of such an agriculture is to preserve and
pass on a healthy soil, the most obvious and directly observable benefit
may be a reduction in the fossil energy required for field work. The top
ten grain, forage, and fiber (cotton) crops currently grown on the 316
million acres considered here now require, for traction in the field, the
equivalent of 71.48 million barrels of petroleum.[8] This amounts to about
a four-day supply for the United States at the 1980 level of petroleum consumption. With elimination of the need for annual preparation of the seed
bed, planting, and cultivation and for fertilizers and pesticides, and with
annual harvest costs remaining fixed, we can expect a substantial savings
in traction energy, perhaps half or more.

It is likely, at least in the early stages of perennial polyculture development, that the fields will have to be plowed and reseeded every few years.
The annual traction energy saved, therefore, depends on the percentage
of the 316 million acres that does not have to be plowed, planted, and
cultivated in any given year. We would still have to spend energy at harvest time. (During harvest there could be a minimal increase in energy

cost; we will have to separate the polyculture seed mix.) Since preparing the seed bed, planting, cultivating, and applying petrochemicals on the 316 million acres accounts for 45 million barrels of petroleum at the wellhead, the annual energy savings is found by multiplying the percentage of the 316 million acres that is not plowed by 45 million barrels. If we replant every five years, 253 million acres would remain undisturbed each year and only 63 million would be disturbed. Fifty-one percent of the current traction energy could thus be saved—the energy equivalent of 36.5 million wellhead barrels.

Energy worth 119 million wellhead barrels is currently applied to the 316 million acres in the form of commercial fertilizers each year.[9] Perennial grain polycultures could fertilize themselves for at least three reasons: (1) the polycultures would contain legumes and some grasses that supply nitrogen, (2) grains that are perennial grasses in the polycultures would suppress nitrification by release of inhibitors from the rhizomes, and (3) perennial polycultures do a better job of capturing and holding rain and snow and thereby make better allocations of the natural nitrogen contained in the water cycle. We would still need about 12 million barrels of petroleum equivalent to replace phosphorus and potassium—a net energy savings equivalent to 107 million barrels.

The most substantial savings would be those of soil, fertility, and traction energy. Some additional savings of lesser importance are still worth mentioning. For example, energy to manufacture equipment would be reduced, a savings that is important to the reduction in equipment use overall. For a three-year replanting cycle, we can show an annual savings of sixteen million wellhead barrels; for a twenty-year replanting cycle, annual savings would be twenty-three million.[10]

The energy savings for reducing pesticide use to zero would not be substantial. All the pesticides applied to the top ten crops growing over the 316 million acres had an energy value equivalent to around 11.5 million barrels at the wellhead.[11] This amounts to only 0.04 percent of 1981 oil consumption. (It is difficult to find a suitable substitute for cotton fiber, which now consumes around one-third of the total energy value of these pesticides. It is perhaps wishful thinking, but if we wore more wool and rotated cotton fields to sheep pasture, there could be substantial savings.) The major benefit from decreased pesticide use is not in the energy saved in manufacture of the products or in field application. The benefit lies in reduced contamination of the countryside.

The energy required to irrigate the 316 million acres of cropland discussed here amounts to 42.2 million wellhead barrels of oil, which is 72 percent of the irrigation energy used on all U.S. cropland. Perennial polyculture, because of increased efficiency of water use, could eliminate the need to mine fossil water for irrigation and thus save energy as calculated here.[12]

The groundwater mining (annual overdraft) for irrigation in the United States is not known, but a rough estimate can be obtained from the fact that the groundwater mining for all water uses in the United States in 1975 was 23.3 million acre-feet and that groundwater for irrigation is 68 percent by volume of groundwater for all water uses in this country. Thus, groundwater mining for irrigation is roughly 15.8 million acre-feet annually. On the average in the United States, it takes 1.01 wellhead barrels of oil to pump one acre-foot of water; thus, 16.0 million wellhead barrels of oil are used annually for groundwater mining for irrigation in the United States. Since we are dealing with the 316 million acres, which accounted for 72 percent of all U.S. on-farm irrigation, then proportionally, the 316 million acres required 11.5 million wellhead barrels of oil for groundwater mining for irrigation.[13]

We know it is unrealistic and unnecessary to stop all irrigating in the United States, but if perennial polyculture, with its efficient use of water, could reduce the need to mine *fossil* water for irrigation, not only would it prevent an annual overdraft of about 15.8 million acre-feet but also the energy savings would be equivalent to 11.5 million wellhead barrels of oil. Mining energy to obtain water to grow food for cattle and pigs that will convert high-quality grain into meat protein in the range of 10–20 percent efficiency is questionable resource management.

If one is looking at yield alone, mixed perennial grain crops may be only marginally competitive, especially in the early stages of development. If, however, we were to allocate no more fossil fuel to conventional agriculture than would be needed in perennial polyculture, we strongly doubt that conventional agriculture would compete with what we envision.

Because of its chemical diversity, a domestic polyculture, like a natural ecosystem, will be less vulnerable to insects and pathogens. The more efficient retention and use of soil moisture by perennial roots makes the domestic ecosystem more resistant to drought than an annual monoculture. It is obviously more resistant to heavy rains as well, and reduced soil loss means retained fertility.

16 Good, Wild, Sacred

Gary Snyder

I live on land in the Sierra Nevada of Alta California, continent of Turtle Island, which is somewhat wild and not terribly good. The indigenous people there, the Nisenan or Southern Maidu, were almost entirely displaced or destroyed during the first decade of the gold rush. Consequently, we have no one to teach us which parts of that landscape were once thought to be sacred, but with much time and attention, I think we will be able to identify such sites again. Wild land, sacred land, good land. At home developing our mountain farmstead, in town at political meetings, and farther afield studying the problems of indigenous peoples, I hear each of these terms emerging. By examining these three categories, perhaps we can get some further insights into the problems of rural habitation, subsistence living, wilderness preservation, and Third and Fourth World resistance to the appetites of industrial civilization.

Wild refers to all unmanipulated, unmanaged natural habitat. Most of the planet in precivilized times was hospitable to humans—rich rainforests, teeming seacoasts, or grasslands covered with bison, mammoths, or pronghorns. Near climax, high biomass, perennially productive, such places were essential expressions of biological nature. Some parts are better than others in terms of supporting much life, with soils rich in nutrients, but even inhospitable mountain terrain may provide special plants or animals of unique value. Knowledge is the real key: for a Kalahari bushman, a Pintubi of the west-central Australian desert, or a Ute of the Great Basin, those arid lands are a life-sustaining home. Many if not all archaic and nonliterate peoples have also found some parts of the landscape to be special, "sacred," and have given etiquette and lore to that. Such spots are of course also wild.

The idea of good land really comes from agriculture. Here *good* is narrowed to mean land productive of a much smaller range of favored cultivars, and thus the opposite of wild, cultivated. In wild nature there is no disorder: no plant in the almost endless mosaics of micro and macro communities is really out of place. For hunting and gathering peoples who draw on that spread of richness, a cultivated patch of land might seem bizarre, and not particularly good, at least at first. Gathering peoples gather from the whole field, ranging widely daily. Agricultural people live by an inner map made up of highly productive nodes (cleared fields) connected by lines (trails through the scary forest). A beginning of "linear."

In civilized agrarian states the term *sacred* was sometimes applied to ritually cultivated land or special temple fields. The fertility religions of those times were not necessarily rejoicing in the fertility of all nature, but were focusing on crops. The concept of cultivation was extended to describe a kind of training in lore and manners that guarantees membership in an elite class. By the metaphor of "spiritual cultivation," a holy man is one who has weeded out the wild from his nature. But weeding out the wild from the natures of members of the Bos and Sus clans—cattle and pigs—tranformed animals that are intelligent and interesting in the wild into sluggish meat-making machines. Cultivation at the top makes domestication and exploitation possible below.

Wild groves and grottoes lingered on as shrines in agrarian states and were viewed with much ambivalence by the rulers from the metropole. They survived because the people who actually worked the land still half-heard the call of the old ways, and certain folk teachings were still being transmitted that went back to even before agriculture. The kings of Israel began to cut down the sacred groves, and the Christians finished the job.

The thought that wild might also be sacred returned to the Occident only with the Romantic movement. This reappreciation of nature projects a rather vague sense of the sacred, however. It is only from very old place-centered cultures that we hear of sacred groves, sacred land, in a context of genuine belief and practice.

In North America and Australia, the original inhabitants are facing the latest round of incursions into their remotest territories. These reservations or reserves were left in their use because the dominant society thought the arctic tundra or arid desert "no good." People of Australia, Alaska, and elsewhere are vigorously fighting to keep logging or oil exploration or uranium mining out of some of their landscapes, and not only for the reason that it is actually their own land but also because some places in it are sacred.

So a very cogent and current political issue rises around the question of the possible sacredness of certain spots. I was at the University of Montana in the spring of 1982 on a program with Russell Means, the American Indian Movement founder and activist, who was trying to get support for the Yellow Thunder Camp of Lakota and other Indian people of the Black Hills on what is currently called Forest Service land. These Indians wish to block further expansion of mining into the Black Hills. They argue that the particular place they are on is not only ancestral land but sacred.

During his term, former California governor Jerry Brown created the Native American Heritage Commission specifically for California Indians, and the commission identified a number of Indian Elders who were charged with the task of locating and protecting sacred sites and graves in California. This would avoid in advance confrontations between landowners or public land managers. It was a sensitive move, and though barely comprehensible to the white voters, it sent a ripple of appreciation through all the native communities. The white Christian founders of the United States were probably not considering American Indian religions when they guaranteed freedom of religion, but interpretations by the courts, and the passage of the American Indian Religious Freedom Act of 1978, have gradually come to give native practices some real status. Sacred virtually becomes a new land-use category.

In the hunting and gathering way of life, the whole territory of a given group is fairly equally experienced by everyone. It becomes known for its many plant communities, high and low terrain, good views, odd-shaped rocks, dangerous spots, and places made special by myth or story. There are places where women go for seclusion or to give birth, places the bodies of the dead are taken to. There are spots where young girls or young boys are called to for special instruction. Some places in this territory are recognized as numinous, loaded with meaning and power. This has happened to all of us. The memories of such spots are very long.

I was in Australia in the fall of 1981 at the invitation of the Australian Aboriginal Arts Board doing some teaching, poetry readings, and workshops with aboriginal leaders and children. Much of the time I was in the central Australian desert south and west of Alice Springs, first into Pitjantjara tribal territory, and then 300 miles northwest into Pintubi tribal territory. The aboriginal people in the central desert all still speak their languages. Their religion is fairly intact, and most young men are still initiated at fourteen, even the ones who go to high school at Alice Springs. They leave the high school with the cooperation of the school authorities for a year, and are taken out into the bush to learn bush ways on foot, to master the lore of landscapes and plants and animals, and finally to undergo initiation.

I was traveling by truck over dirt track west from Alice Springs in the company of a Pintubi elder named Jimmy Tjungurrayi. As we rolled along the dusty road, sitting in the bed of a pickup, he began to speak very rapidly to me. He was talking about a mountain over there, telling me a story about some wallabies that came to that mountain in the dreamtime and got into some kind of mischief there with some lizard girls. He had hardly finished that and he started in on another story about another hill over here and another story over there. I couldn't keep up. I realized after about half an hour of this that these were tales to be told while *walking*, and that I was experiencing a speeded-up version of what might be leisurely told over several days of foot-travel. Mr. Tjungurrayi felt graciously compelled to share a body of lore with me by virtue simply of the fact that I was there.

So remember a time when you journeyed on foot over hundreds of miles, walking fast and often traveling at night, traveling night-long and napping in the acacia shade during the day, and these stories were told to you as you went. In your travels with an older person you were given a map you could memorize, full of the lore and song, and also practical information. Off by yourself you could sing those songs to bring yourself back. And you could maybe travel to a place that you'd never been, steering only by songs you had learned.

We made camp at a waterhole called Ilpili and rendezvoused with a number of Pintubi people from the surrounding desert country. The Ilpili waterhole is about a yard across, six inches deep, in a little swale of bush full of finch. The people camp a quarter mile away. It's the only waterhole that stays full through drought years in several thousand square miles. A place kept by custom, I am told, welcome and open to all. Through the night, until one or two in the morning, Jimmy Tjungurrayi and the other old men sat and sang a cycle of journey songs, walking through a space of desert in imagination and song. They stopped between songs and would hum a phrase or two and then would argue a bit about the words and then would start again, and someone would defer to another person and would let him start. Jimmy explained to me that they have so many cycles of journey songs they can't quite remember them all, and that they have to be constantly rehearsing them. Night after night they say, "What will we sing tonight?" "Let's sing the walk up to Darwin." They'll start out and argue their way along through it, and stop when it gets too late to go any farther. I asked Jimmy, "Well, how far did you get last night?" He said, "Well, we got two-thirds of the way to Darwin." This is a way to transmit information about vast terrain that is obviously very effective and doesn't require writing. Some of the places thus defined will also be presented as sacred.

One day driving near Ilpili we stopped the truck and Jimmy and three other elderly gentlemen got out and said, "We'll take you out to see a sacred place here." And, "I guess you're old enough." They turned to the young boys and said that uninitiated boys couldn't go there. As we climbed the hill, these ordinarily cheery and loud-talking aboriginal men began to drop their voices. As we got higher up the hill, they were speaking in whispers, their whole manner changed. They said, in a whisper, "Now we are coming close." Then they got on their hands and knees and crawled. We crawled up the last 200 feet, over a little rise into an area of broken and oddly shaped rocks. They whispered to us with respect and awe of what was there and its story. Then we all backed away. We got back down the hill and at a certain point stood and walked. At another point voices rose. Back at the truck, everybody was talking loud again and no more mention was made of the sacred place.

Very powerful. Very much in mind. We learned later that it was a place where young men were taken for instruction and for initiation.

So the nature of the "sacred place" in Australia began to define itself as special rocks, beautiful, steep defiles where two cliffs almost meet with maybe just a little sand bed between, a place where many parrots are nesting in the rock walls, or a place where a blade of rock stands on end balancing, thirty feet tall, by a waterhole. Each of them was out of the ordinary, a little fantastic even, and they were places of teaching. Often they had pictographs, left by past human ancestors. In some cases they were also what are called "dreaming spots" for certain totem ancestors. "Dreaming" or "dreamtime" refers to a time of creation which is not in the past but which is here right now. It's the mode of eternally creative nowness, as contrasted with the mode of cause and effect in time, where modern people mainly live and within which we imagine history, progress, evolution to take place. The totem dreaming place is first of all special to the people of that totem, who sometimes make pilgrimages there. Second, it is sacred to the honey-ants (say) that actually live there. There are a lot of honey-ants there. Third, it's like a little Platonic cave of ideal honey-ant forms. (I'm imagining this now. I'm trying to explain what all these things seem to be.)

It's the archetypal honey-ant spot. In fact, it's optimal honey-ant habitat. A green parrot dreaming place, with the tracks of the ancestors going across the landscape and stopping at the green parrot dreaming place, is a perfect green parrot nesting spot. So the sacredness comes together with a sense of optimal habitat of certain kinfolk that we have out there—the wallabies, red kangaroo, bush turkeys, lizards. Robert Bliney sums it up this way: "The land itself was their chapel and their shrines were hills and creeks and their religious relics were animals, plants, and birds. Thus the migrations of aboriginals, though spurred by economic need, were also always pilgrimages." Good (productive of much life), wild (naturally), and in these cases, sacred, were indeed one.

This way of life is going on right now, threatened by Japanese and other uranium mining, large-scale copper mining, and petroleum exploration throughout the deserts. The issue of sacredness is a very real political question, so much so that the Australian Bureau of Aboriginal Affairs has hired some bilingual anthropologists and bush people to work with elders of the differènt tribes to identify sacred sites and map them. Everyone hopes that the Australian government really means to declare such areas off-limits before any exploratory team ever gets near them. This effort is spurred by the fact that there have already been some confrontations in the Kimberley region over oil exploration. This was at Nincoomba. The people very firmly stood their ground and made human lines in the front of bulldozers and drilling rigs and won the support of the Australian public. Since then the Australian government has been more careful. In Australian land ownership, mineral rights are always reserved to "the crown" so that even a private ranch is subject to mining. To consider sacred land a special category in Australia is a very advanced move, at least in theory. But recently a "registered sacred site" was bulldozed near Alice Springs, supposedly on instructions of a government land minister, and this is in the relatively benign federal government jurisdiction. The state of Queensland is a mini-fascist nation to itself, favored by emigrants from white South Africa.

The original inhabitants of Japan, the Ainu, can see a whole system as in a very special sense sacred. Their term *iworu* means "field" with implications of watershed, plant and animal life, and spirit force. They speak of the *iworu* of the great brown bear. By that they mean the mountain habitat and watershed territory in which brown bear is dominant. They also speak of the *iworu* of the salmon, which means the lower watersheds with all their tributaries and the plant communities along those valleys that focus on the streams where salmon run. The bear field, the deer field, the salmon field, the orca (killer whale) field. To give a little picture of how this world works, a human house is up a valley by a stream, facing east. In the center of the house is the fire. The sunshine streams through the eastern door each morning to contact the fire, and they say the sun goddess is visiting her sister the fire goddess in the firepit. They communicate for a moment. One must not step across the sunbeams that shine in the morning on the firepit; that would be breaking their contact.

Food comes from the inner mountains and from the deeps of the sea. The lord of the deeps of the sea is Orca or Killer Whale, the lord of the inner mountains is Bear. Bear sends his friends the deer down to visit us. Killer Whale sends his friends the salmon up the streams to visit us. When they come to visit us we kill them, to enable them to get out of their fur or scale coats, and then we entertain them because they love music. We sing songs to them, and we eat them. Having been delighted by the songs they heard, they return to the deep sea and to the inner mountains, and they report to their spirit friends there, "We had a wonderful time with the human beings. There's lots to eat, lots to drink, and they played music for us." The other ones say, "Oh, let's go visit the human beings." If the people do not neglect the proper hospitality, the music and manners, when entertaining their deer or salmon or wild plant-food visitors, the beings will be reborn and return over and over. This is a sort of spiritual game management.

The Ainu were probably the original inhabitants of all of Japan. They certainly left many place-names behind and many traces on the landscape. Modern Japan is another sort of example: a successful industrialized country, with remnants of sacred land-consciousness still intact. There are Shinto shrines throughout Japan. Shinto is "the way of the spirits." By *spirits* the Japanese mean exactly what almost all people of the world have always meant: spirits are formless little powers present in everything to some degree but intensified in power and in presence in outstanding objects, such as large curiously twisted rocks, very old trees, or thundering misty waterfalls. Anomalies and beauties of the landscape are all signs of *kami*—spirit power, spirit presence, energy. The greatest of all the *kami*, or spirit forces of Japan, is Mt. Fuji. The name Fuji is now thought to be an old Ainu place-name meaning "fire goddess." All of Mt. Fuji is a Shinto shrine, the largest in the nation, from well below timberline all the way to the summit.

Shinto got a bad name during the 1930s and World War II because the Japanese government created a "State Shinto" in the service of militarism and nationalism. Long before the rise of any state, the islands of Japan were studded with little shrines—*jinja* or *miya*—part of the expression of Neolithic village culture. Even in the midst of the enormous onrushing industrial energy of the current system, shrine lands remain untouchable. It would make your hair stand up to see how the Japanese will take bulldozers to a nice slope of pines and level it for a new development. When the

New Island was created in Kobe harbor, to make Kobe the second busiest port in the world (next to Rotterdam), it was raised from the bay bottom with dirt obtained by shaving down a range of hills ten miles south of the city. This was barged to the site for twelve years, a steady stream of barges carrying dirt off giant conveyor belts, totally removing soil two ranges back from the coast. That leveled area was then used for a housing development. In the industrial world it's not that "nothing is sacred," it's that the sacred is sacred and that's *all* that's sacred. We are grateful for the little bit of Japanese salvaged land because the rule in shrine lands is that (away from the buildings and paths) you never cut anything, never maintain anything, never clear or thin anything. No hunting, no fishing, no thinning, no burning, no stopping of burning.

Thus pockets of climax forests here and there, right inside the city, and one can walk into a shrine and be in the presence of an 800-year-old cryptomeria tree. Without shrines we wouldn't know so well what Japanese forests might have been. But such compartmentalization is not healthy: in this model some land is saved, like a virgin priestess, some is overworked endlessly like a wife, and some is brutally publicly reshaped, like an exuberant girl declared promiscuous and punished. Good, wild, and sacred couldn't be farther apart.

Europe and the Middle East inherit from Neolithic and Paleolithic times many shrines. The most sacred spot of all Europe was perhaps the caves of southern France, in the Pyrenees. We shall say that they were the great shrines of 20,000 years ago, the center of a religious complex in which the animals were brought underground. Maybe a dreaming place. Maybe a thought that the archetypal animal forms were thereby stored under the earth, a way of keeping animals from becoming extinct. But many species did become extinct. Most became so during the past 2,000 years, victims of the imperium, of civilization, in its particularly destructive western form. The degradation of wild habitat and extinction of species, the impoverishment and enslavement of rural people and subsistence economies, and the burning alive of nature-worship traditions were perfected right within Europe.

So the French and English explorers of North America and then the early fur traders and hunters had no traditions from the cultures they left behind that would urge them to look on wild land with reverence. They did find much that was awe inspiring; some joined the Indians and the land and became people of place. These few almost forgotten exceptions were overwhelmed by fur-trade entrepreneurs and, later, farmers. Yet many kept joing the Indians in fact or in style—grieving for a wilderness they saw shrinking away. In the Far East, or Europe, a climax forest or prairie, and all the splendid creatures that live there, are tales from the Neolithic. In the western United States it was our grandmothers' world. For many of us, without intellectualization or question, this loss is a source of grief. For Native Americans this loss is a loss of land, life, and culture.

It is of course not evil to, as Thoreau did, "make the soil say beans"—to cause it to be productive to our own notion—but we must also ask, what does mother nature do best here when left to her own long strategies? This comes to asking, what would the climax vegetation of this spot be? For all land, however long wasted and exploited, if left to nature, the *tzuran*, "self-so" of Taoism, will arrive at a point of balance between biological productivity and stability. A truly sophisticated postindustrial "future primitive" agriculture will be asking: is there any way we can go with rather than against a natural tendency toward, say, deciduous hardwoods—or as where I live, a mix of pine and oak with kitkitdizze ground cover? Such a condition in many cases might be best for human interest too, and even in the short run.

Wes Jackson's research indicates that a perennial and horticultural-based agriculture holds real promise for sustaining the locally appropriate communities of the future. This is acknowledging that the source of fertility ultimately is the "wild." It has been said that "good soil is good because of the wildness in it." How could this be granted by a victorious king dividing up his spoils? (Spanish land grants—royal/real estate?) In my imagination the God/dess that gives us land is no other than Gaia herself: the whole network.

It might be that almost all civilized agriculture has been on the wrong path from the beginning, relying on the relative monoculture of annuals. In *New Roots for Agriculture*, Wes Jackson develops this argument. I concur with his view, knowing that it raises even larger questions about civilization itself, a critique I have worked at elsewhere. Suffice it to say that the sorts of economic and social organization we invoke when we say "civilization" can no longer be automatically accepted as useful models. To scrutinize civilization as Dr. Stanley Diamond has in "In Search of the Primitive" is not, however, to negate all varieties of culture or cultivation.

The word *cultivation* in civilization, harking to etymologies of *till* and *wheel about*, generally implies a movement away from natural process. Both materially and psychologically, it is a matter of "arresting succession, establishing monoculture." Applied on the spiritual plane, this has meant austerities, obedience to religious authority, long bookish scholarship, or a dualistic devotionalism (sharply distinguishing "creature and creator") and an overriding metaphor of divinity being "centralized," just as a secular ruler of a civilized state is at the center—of wealth, of the metropole, of political power. A divine king. The efforts entailed in such a spiritual practice are sometimes a sort of war against nature—placing the human over the animal, the "spiritual" over the human. The most sophisticated modern variety of this sort of thought is found in the works of Father Teilhard de Chardin, who claims a special evolutionary spiritual destiny for humanity under the name of higher consciousness. Some of the more extreme of these Spiritual Darwinists would willingly leave the rest of earth-bound animal and plant life behind to enter a realm transcending biology. The anthropocentrism of some New Age thinkers is countered by the radical critique of the deep ecology movement.

Yet there is such a thing as training. The natural world moves by process and by complementarities of young and old, foolish and wise, ripe or green, raw or cooked. Animals too learn self-discipline and caution in the face of desire and availability. There are learning and training that go with the grain of things. In early Chinese Taoism, "training" did not mean to cultivate the wildness out of oneself, but to do away with arbitrary and delusive conditioning—false social values distorting an essentially free and correct human nature. Buddhism takes a middle way, allowing as how greed, hatred, and stupidity are part of the given conditions of human nature, but seeing organized society, civilization, "the world" as being a force that inflames, panders to, or exploits these weaknesses in the fledgling human. Greed exposes the foolish person or the foolish chicken alike to the ever-watchful hawk of the food-web, and to early impermanence. Preliterate hunting and gathering cultures lived well by virtue of knowledge and a quiet sort of manipulation of systems. We know how the people of Mesolithic Britain selectively cleared or burned, in the valley of the Thames, as a way to encourage the growth of hazel. An almost invisible horticulture was once practiced in the jungles of Guatemala. The spiritual equivalent of nature-enhancing practices can be seen in those shamanistic disciplines which open the neophyte's mind to that fascinating wild territory, the unconscious.

We can all agree: there is a problem with the chaotic, self-seeking human ego. Is it a mirror of the wild and of nature? I think not: for civilization itself is ego gone to seed and institutionalized in the form of the state, both Eastern and Western. It is not nature-as-chaos that threatens us (for nature is orderly) but ignorance of the real natural world, the myth of progress, and the presumption of the state that it has created order. That sort of "order" is an elaborate rationalization of the greed of a few.

Now we can look again at what sacred land might be. For a people of an old culture, all their mutually owned territory holds numinous life and spirit. Certain spots are of high spiritual density because of their perceived animal or plant habitat peculiarities, or associations with legend and perhaps with human ancestry via totemic systems, or because of their geo-morphological anomaly and formal intensity, or because of their associa-tion with spiritual training, or some combination of these features. These spots are seen as points on the landscape at which one can more easily en-ter a larger-than-human, larger-than-personal, realm.

Nowadays some present-day inhabitants of Turtle Island, and many Eu-ropeans, join with the native peoples of the world in a rather new political and economic movement concerned with "the ecology." Stephen Fox says it is also probably a new religion, so new that it has not been called such yet. Though sometimes attacked as being an elitist movement (even by the Reagan administration), the growing popularity of the Earth First! organi-zation and its "Rednecks for Wilderness" bumpersticker in blue-collar areas shows this to be not true. The temples of this movement are the planet's remaining wilderness areas. When we enter them on foot we can sense that the *kami* or (Maidu) *kukini* have fled here for refuge, as have the mountain lions, mountain sheep, and grizzlies. (Those three North Ameri-can animals were found throughout the lower hills and plains in prewhite times.) The rocky icy grandeur of the high country reminds us of the over-arching wild systems that nourish us all—even an industrial economy, for in the sterile beauty of mountain snowfields and glaciers begin the little streams that water the huge agribusiness fields of the San Joaquin Valley of California. The backpacker-pilgrim's step-by-step, breath-by-breath walk up a trail, carrying all on the back, is so ancient a set of gestures as to trig-ger perennial images and a profound sense of body mind joy.

Not just backpackers, of course. The same happens to those who sail in the ocean, kayak rivers, tend a garden, even sit on a meditation cushion. The point is in making intimate contact with wild world, wild self. *Sacred* refers to that which helps take us out of our little selves into the larger self of the whole universe.

Inspiration, exaltation, insight do not end, however, when one steps out-side the doors of the church. The wilderness as a temple is only a begin-ning. That is: one should not dwell in the specialness of the extraordinary experience, not leave the political world behind to be in a state of height-ened insight. The best purpose of such studies and backpack hikes is to be able to come back into the present world to see all the land about us, agri-cultural, suburban, urban, as part of the same giant realm of processes and beings—never totally ruined, never completely unnatural. Great Brown Bear is walking with us, salmon swimming upstream with us, as we stroll a city street.

To return to my own situation: the land my family and I live on in the Sierra Nevada of California is "barely good" from an economic standpoint. With soil amendments, much labor, and the development of ponds for watering, it is producing a few vegetables and some good apples. As forest soils go it is better: through the millennia it has excelled at growing oak and pine trees. I guess I should admit that it's better left wild. It's being "managed for wild" right now—the pines are getting large again and some of the oaks were growing here before a white man set foot anywhere in California. The deer and all the other animals move through with the exception of grizzly bear; grizzlies are now extinct in California. We dream sometimes of trying to bring them back.

These foothill ridges are not striking in any special way, no great scenery or rocks—but the deer are so at home here, I think it might be a "deer field." And the fact that my neighbors and I and all of our children have learned so much by taking our place in the Sierra foothills—not striking wilderness, but logged-over land, burned-over land, considered worthless for decades—begins to make it a teacher to us. A place on Earth we work with, struggle with, where we stick out the summers and winters. And it has showed us a little of its power.

But this use of "teacher" is still a newcomer's metaphor. By our grandchildren's time there may begin to be a culture of place again in America. How does this work? First, a child must experience that bonding to place that has always touched many of us deeply: a small personal territory one can run to, a secret "fort," a place of never-forgotten smells and sounds, a refuge away from home. Second, one must continue to live in a place, to not move away, and to continue walking the paths and roads. A child's walking the land is a veritable exercise in "expanding consciousness." Third, one must have human teachers, who can name and explain the plants, who know the life cycle of an area. Fourth, one must draw some little part of one's livelihood from the breadth of the landscape: spotting downed trees for next year's firewood, gathering mushrooms or berries or herbs on time, fishing, hunting, scrounging. Fifth, one must learn to listen. Then the voice can be heard. The nature spirits are never dead, they are alive under our feet, over our heads, all around us, ready to speak when we are silent and centered. So what is this "voice"? Just the cry of a flicker, or coyote, or jay, or wind in a tree, or acorn whack on a garage roof. Nothing mysterious, but now you're home.

Fine, and what about right now? As Peter Nabokov says, good-hearted environmentalists can turn their back on a save-the-wilderness project when it gets too tiresome and return to a city home. But inhabitory people, he says, will "fight for their lives like they've been jumped in an alley." Like it or not, we are all finally "inhabitory" on this one small blue-green planet. It's the only one with comfortable temperatures, good air and water, and a wealth of living beings for millions (or quadrillions) of miles. A little waterhole in the vast space, a nesting place, a place of singing and practice, a place of dreaming. It's on the verge of being totally trashed— there's a slow way and a fast way. It's clearly time to put hegemonial controversies aside, to turn away from economies that demand constant exploitation of both people and resources, and to put Earth first.

As the most numerous, ambitious, and "musical" (as the Ainu would say) sort of the larger mammals, human beings might well awaken to their great possible place in the biosphere as sensitive transformers. We might someday initiate a more sophisticated dialogue between the poles of cultivation and original nature, technology and the self-born, production and reproduction, than has ever been imagined before. These possibilities go far beyond any fantasies of high-tech. I'm thinking of a condition where wild, sacred, and good will be one and the same, again.

17

A Search for the Unifying Concept for Sustainable Agriculture

Wes Jackson

The use of "*the*"—instead of "*a*"—"unifying concept" in the title of this chapter was not accidental. I chose it because I believe that a truly sustainable agriculture will be directly keyed to nature, which already has a well-understood unifying concept of its own. I am speaking of the unifying concept of biology as discovered by Darwin and which was later coupled with the discovery that the DNA-RNA hereditary code is universal. Essentially all the natural ecosystems of the earth are many times more complex than our most elaborate agricultural ecosystems, and if there can be a unifying principle for the diverse natural biota, why shouldn't there be one for agriculture, especially if nature is our standard?

Some could argue, of course, that agriculture is so much a product of human manipulation that a unifying concept will have to be *invented*. I don't think so. I think it is only to be *discovered*. There are enough examples of good farming that we recognize as being in harmony with nature that we can look to them. By good farming, I mean practices that don't consume ecological capital or otherwise degrade the landscape. That such examples are few in number doesn't matter. Other possibilities for agriculture that are theoretically possible, although yet to be developed, take nature's designs even more into account than the best agricultural examples now available.

This search did not begin with this chapter. It is at least as old as the Hebrew scriptures and has been getting more complicated ever since. Before the industrial age, soil loss alone was at the core of the "problem of agriculture." It is currently the oldest of our ecological problems, for it probably has its origin with the first tilled crop laid out on a sloping landscape. The industrial revolution added other problems to the core—fossil-fuel dependency for traction and fertility and chemical dependency for pest control. The problem becomes even more complicated once we add the structural changes that have especially influenced agriculture in the developed world as the result of new technologies interacting with politics and economics.

THE HIERARCHY OF STRUCTURE

As problems of production agriculture became more complicated, it is probably natural that we would begin to look for a way of thinking about the definition and possibility of sustainable agriculture, a conceptual tool that will provide a perspective on all crops and all management practices. First of all, we need a way of thinking about any agricultural unit: the family garden, the truck farm, the small family farm, the large corporate farms, and the individual field. That is one thing a unifying concept would do. It would be fluid and become the basis for a taxonomy even though taxonomic schemes are not fluid. Concepts, like movies, deal with process. Biological classification, at least, is expressed as single frames, stopped in time. A taxonomic scheme can be totally arbitrary. One could classify agricultural efforts, for example, based on acreage or principle crops or livestock grown, but the utility of such a scheme would be limited unless certain laws about "the nature of things" stood behind the classification system.

Professional ecologists do research and talk about the possibilities and limits of natural diversity. Why should we not be able to deal with farm ecosystems in a similar manner? What if population biologists considered how populations of organisms interact with other populations on the farm as well as with the physical and chemical world? If the taxonomy took into account these considerations and the laws that are in operation as new wild species come into existence, the classification would surely be regarded as more natural *and* more complete. The concept would provide the basis for a taxonomy by making us more conscious of the variables that must be adjusted if agriculture is to be sustainable in any given place. The best taxonomic scheme for our purposes, then, is more than an arbitrary system of classification, for it would be embedded in a theory of structure.

I don't mean to imply that we will ever have a perfect taxonomy which shows precisely where and how everything fits. That would be asking too much. Our current classification of the biota is only an approximation of the phylogenetic relationships in the tree of life, and a once-and-for-all classification of nature's life forms based on phylogeny still eludes us. But just as Charles Darwin's unifying concept for biology—evolution through natural selection—made a natural classification of our earth's biota theoretically possible, so might a unifying concept for sustainable agriculture make a classification of sustainable agriculture methods *theoretically* possible even though it may be practically very difficult.

In his paper on the ecosystem as a conceptual tool in managing natural resources, Arnold Schultz explains why ecosystems should be studied as a separate science. He thinks the study of ecosystems should be a separate field because it is not like studying botany, zoology, not even ecology. Schultz argues that such a science would provide a framework for analyzing any organization integrated immediately above the level of the individual organism.[1]

The ecologist A. G. Tansley, who coined the term *ecosystem* in 1935, described it as "the whole system, including not only the organism complex, but also the whole complex of physical factors forming what we call the environment."[2] It would be more accurate to say, "forming what we normally have been calling the environment of the organisms." Defining *ecosystem* more succinctly than Tansley, R. L. Lineman described it as "a system composed of physical-chemical-biological processes active within a space-time unit of any magnitude."[3]

Good descriptions are difficult. Both of these could, I suppose, include the earth, sun, and moon, and so they are limited. Rather than worry for now how large an ecosystem can be before it is no longer an ecosystem, let us consider where it stands in relation to everything else downward in the universe. J.S. Rowe says the ecosystem is part of an integrated hierarchy of things from atoms to molecules to cells to tissues to organs to organisms.[4] The ecosystem stands immediately above the organism. That the ecosystem should stand at this particular place is not immediately intuitive because beyond organisms we are used to thinking about plant communities, biotic communities, vegetation, populations, species, and world fauna and flora. Schultz asked this question and finally agreed with Rowe that what an ecosystem has which these categories do not have is "thinghood." Rowe's argument is that nature has "chunks of space-time or 'events' which have both qualitative and quantitative properties. Events which endure are known as objects." For something to have "thinghood"—that is, to be an object—it must have volume because volume is the basic component of perception. These other categories, species, populations, etcetera, lack volume because they do not include their surroundings.

But volume alone won't do because, as Schultz says (leaning on Rowe), "for objects to have a high degree of 'thinghood' they must exist in both space and time. Form and function must be constant or have rhythmic stability. Organized entities have strongly marked structure and function. When organized entities have strongly marked structural and functional characteristics, they are *perceived as autonomous* and stand out as natural objects of study" (my italics). The integrative level of atom, molecule, cellular organelles, cell, tissue, organ, organ systems, and organism are all natural objects of study. They are "slabs of space-time," as Rowe calls them. The volumetric criterion holds for them all. Before ecosystem was added to the hierarchy, J. K. Feibleman wrote twelve laws of integrative levels.[5] With this volumetric consideration, Rowe rewrote one of Feibleman's laws so that volume and space relationships are included. This law holds that *the object of study, at any level whatever, must contain, in the volume sense, the objects of the lower level, and must itself be a volumetric part of the levels above.* Each object on any given level constitutes the immediate environment (in the sense of impinging surroundings) of objects on the level below. Each object is a specific structural and functional part of the object at the level above. The ecosystem passes this test as an object because it consists of individual organisms plus the nonliving world that connects them. (What the next natural "chunk" or "slab" beyond ecosystem in nature is, I can't say. Perhaps it is bioregion or continent or ecosphere.) What is important for agriculturists is that Rowe seems to have made the case that the next integrative level above organism is the ecosystem as the space-time unit. To repeat, species, population, vegetation, or community are not "environment" for individual plants, and they are not any specific volumetric functional part of the ecosystem.

The ecosystem does differ from the other categories in the hierarchy because the human defines the boundary. The boundary of an organism is natural and well understood. The same is true of an organ or a cell. There are certain natural ecosystems, such as bogs, in which the boundary is clear. However, for most natural ecosystems, it is difficult to know, with much precision, where the boundary lines are. Particularly problematic is knowing where tall-grass prairie ends and mid-grass prairie begins or where mid-grass prairie ends and short-grass prairie begins. Yet few prairie ecologists would deny the existence of the three prairie types.

With agricultural ecosystems, where we place the boundary is more suggested by a human-imposed pattern on the landscape than by nature. A
deed can define the boundary of a farm. A fence line may define a pasture
or a field. We could think about a farm community as an ecosystem or a
farm or an alfalfa field or even a cubic meter of soil. We can place mental
cubes around anything we want because our purpose is to be better ac-
countants of what goes in and out through the boundary and, at the same
time, appreciate the dynamics of the ecosystem as a structure that obeys
certain laws common to the other levels in the hierarchy.

LAWS OF THE INTEGRATIVE LEVELS

Before we dwell on the ecosystem level as an integrative level, we need to
consider the other laws of integrative levels distilled by Schultz to apply to
the ecosystem. Schultz has also derived these other laws and their corollaries from Feibleman's twelve laws. Next, we can't ignore species and
populations if we are to talk about agriculture because we can't ignore
populations of cows or wheat. Even though species and populations did
not attain the status of being part of the hierarchy of structure that
Feibleman, Rowe, and Schultz have discovered, they were regarded seriously as candidates. They just failed the test of "thinghood." They are,
nevertheless, part of another kind of hierarchy, a hierarchy of biological
descent. This hierarchy has its own laws, and species and populations exhibit some dynamic properties fundamentally different from what individual organisms do. They generate species diversity, and species adaptation
is of a different order than individual adaptation. Knowledge of these dynamic properties and the ways in which they impinge on the ecosystem
will be useful in the effort to discover the unifying concept for sustainable
agriculture. But first let us look at the other laws of integrative levels that
Arnold Schultz, writing with the ecosystem in mind, developed from J. K.
Feibleman's scheme.

Schultz combined some of the laws and corollaries of Feibleman into
seven categories that he felt were the most relevant to the development of
the ecosystem concept. I have already mentioned the seventh law, which
deals with the criteria for "thinghood" and makes the ecosystem an integrative level. The first six laws follow, almost as Schultz has presented
them, but not in the same order. I am listing the self-evident laws first.
They are not necessarily trivial, but an elaboration on their relevance to
agricultural ecosystems may not be necessary at this time.

   1. *In any organization, the higher level depends on the lower.* Just as the
organism depends on organs, organs on tissues, tissues on cells, etcetera,

so the ecosystem depends on organisms, soil, and water. What is implied here is that the lower the level is, the more enduring the level is. Atoms are more lasting than molecules. The physical and biological components of an ecosystem are more lasting than the ecosystem.

2. *The higher the level, the smaller the population of instances.* There are fewer molecules than atoms and fewer ecosystems than organisms. The levels form a population pyramid. This law still holds for agricultural ecosystems. The variety of farms in Kansas may be greater than the number of natural ecosystems, but the number of total ecosystems still does not exceed the number of organisms associated with the farms.

3. *Complexity of the levels increases upward.* Complexity is partly the result of accumulating structure, but most of the complexity stems from emergent qualities that pile up. The emergent qualities are the interrelationships that increase exponentially while the number of components increase linearly.

4. *Each level organizes the next level below and adds emergent qualities* (see number 3). If one knows only the properties of the lower level, the emergent qualities are unpredictable. Schultz points out that we couldn't have known from the gases hydrogen and oxygen that they could produce water. In this example more than one water molecule had to exist before the liquid property could emerge.

There are examples at other levels of organization where "critical mass" is necessary before an emergent quality can arise. For example, one cell does not a tissue make. This "critical mass" idea may have great practical importance for sustainable agriculture. There is probably no such thing as a completely sustainable farm anywhere in the United States. Some farms are more resilient or can stand being weaned away from the fossil-fuel economy better than others. The Amish are probably the most notable, as a group, in this respect. But it is rare that an Amishman or Amish family will venture forth, alone, into a locality not previously settled by Amish people. When colonies are founded, seven families usually go together. The Amish know they can't make it in isolation. Although there are probably a lot of isolated good farmers whose farms experience no net soil loss, their ancestors were probably heavily dependent on community during the farms' establishment. A lot of back-to-the-landers learned that most individuals or even individual families who have made attempts at sustainable agriculture have fared poorly or failed outright. Perhaps a "critical mass" is necessary before the emergent qualities necessary for a sustainable agriculture begin to appear. For our purposes, the implication is that sustainable agriculture will need rural communities if it is to survive and flourish. In retrospect, it is clear that, once the systematic destruction or dismantling of rural communities was underway in the United States, the weakening of the family farm was inevitable.

5. *For an organization at any given level, its mechanism lies at the level below and its purpose at the level above.* This law may need to be restated. As it now stands, it presents a particularly sticky problem when thinking about nature. Ironically, it serves us well in thinking about conventional agricultural ecosystems even though it does not apply to natural ecosystems. We may know the purpose of a cornfield, but what is the purpose of tall-grass prairie? Our problem becomes especially difficult if we don't keep the origin of a structure separate from what we finally see as a finished product. For example, if we drop down to the organ level and take the kidney, we might say that the purpose of the kidney is to remove nitrogenous wastes from the blood of an organism—to cleanse it. Here the purpose lies above the organ (at the organism level) and the mechanism lies below the organ (in the tissue). But few biologists would be satisfied with such an explanation, for in the evolution of the vertebrate kidney, few of them would believe that purpose was "pulling" on one end, forcing the development of a mechanism to accommodate a higher purpose. Biologists, instead, speak of adaptation and say that those early creatures that had tissues fortuitously tilted in the direction of removing nitrogenous wastes, however slight, were positively selected, in a Darwinian sense. Creatures that carried improvements on each former adaptation were further selected, and so on.

Only from our vantage point in history do we assign the words *purpose* and *mechanism* for what nature has produced. Even if we had been observers in the early stages of the evolution of the kidney, we probably would have considered the primitive or elementary kidney a finished product.

Perhaps part of the human condition is the result of the distance from nature in which we place ourselves by dealing with the world in a language that emphasizes purpose and mechanism. An example from agriculture is a typical Kansas wheat field. If it is truly representative of a field on a typical Kansas farm, its purpose is to provide cash outright for the farmer. The bottom line, in other words, is production. The farmer needs a high yield. Since his emphasis is on production, he naturally employs mechanisms of "mass production," products of industry. Mass production features a huge capitalization of equipment and inputs and seeks to minimize labor costs. The logical outcome is larger and more expensive machinery and fewer people. But this creates vulnerability. If the crop is to get planted, tended, and harvested, breakdowns of equipment become less tolerable. For many farmers, this often means new equipment every two or three years and a need for higher yields to pay for it. The *purpose* of the usual American farmer, to produce cash, almost to the exclusion of everything else, *dictates the mechanism* for growing wheat.

On 160 acres recently purchased by The Land Institute, there is a clear example of the impact on the land from the purposes of production agriculture. On our quarter are two small streams that run during part of the year but are usually dry by the middle or end of summer. This quarter section had been held in a trust by a local bank for a young man for twenty-five years and had been rented out to one farmer for the past seventeen years. As a farmer who derives his total income from farming, he is typical for our area. He tills around 1,200 acres of land and has a cattle operation, too. Because these two streams and their woody growth proved to be such a nuisance for his large equipment during the dry period when it is time to prepare the ground to plant wheat, he and a trust officer at the bank, probably with government funds, had the riparian community bulldozed out so that there was one field that he could "farm right through." He could lower his production costs if he wasn't slowed down by small fields. Tree roots would rob some nutrients from the wheat along the edge. Most years it is still too wet in June to cut the wheat that grows in this wet spot, but it does not slow him down during seed bed preparation and planting in late summer and early fall. Ironically, when a large limb of a big hackberry growing on the property line fell directly into the field a few years ago, he farmed around the limb. Because he did not have to reduce the speed of his tractor or make a wide turn to farm around the limb, it was, because of his purpose, not cost effective to remove it. Neither was it cost effective to till the ground on the contour or build terraces. It was cost effective to pull into the field and start going around the perimeter of the field and move toward the middle, letting the perimeter, not the streams, dictate the pattern. The living riparian community was bulldozed out, the dead limb was allowed to encroach on the field, and the natural topography was ignored in tilling and planting, all for the same purpose.

I recently visited several Amish farms in Ohio where the topography is much more rugged than in central Kansas. Streams divide their fields, too, but their farms have a very different appearance. What is underway on an Amish farm does not involve single purpose. The farms are not regarded as economic units, although the Amish make sound economic decisions. What we observe on the Amish farms is similar to what we observe on a natural ecosystem—homeostasis. Purpose and mechanism are transcended.

One Amishman told me that he liked the long-stemmed wheat because the straw was about as important to him as the grain. He used the straw for bedding by which he salvaged and improved his manure, and he was surprised to hear that we Kansans would either plow the straw under or burn it. The Amish feed some of the grain to livestock, some they sell. Although they take the Biblical injunction to "dress and keep the earth" seriously, one does not sense that they go about their daily work with that phrase rolling around in their heads. They are interested in profit and high yield, but neither concern drives them as a singular purpose. Had The Land Institute's newly acquired 160 acres been an Amish farm, it would have been highly diversified, and provided there was a surrounding community of Amish, that 160 acres would easily have supported one Amish family. The living riparian community on each side of the two streams would have been a habitat for an abundance of wild species including quail, pheasant, and deer.[6] It would have been a source of fuel, a boundary dividing the farm into smaller fields. It would break the hot, dry wind of summer, interrupt insect migration, and host some predatory birds and insects. The smaller fields would have suited a horse- or mule-powered agriculture. The large cottonwoods would have provided shade for grazing animals or for a resting team and driver. The fallen hackberry limb would have been converted into firewood. The straw that we plow under or burn would have become bedding for livestock and thus become a way of holding urine and manure, and all three would have been returned to the fields from which they came. Some of the grain would be fed on the farm, some would be sold, depending on need.

Because the emphasis for the Amish is not exclusively on production, mass production of food on the farm is incompatible. To "dress and keep the earth" is biology. Mass production is purposeful and invites mechanism. Mechanism and machines go together. That is why conventional farmers are obligated to assume the high capital costs of expensive equipment. To ensure that the crops are planted, protected, and harvested, the Amish depend on a sufficiency of people who live on the land and do the work. For the Amish, *resilience* lies with that sufficiency of people. It isn't likely that during harvest all the workers will catch the summer flu at the same time. Furthermore, the Amish, while always busy, are not pressed for production; if a draft animal does become sick, harvest isn't shut down. Wheat is cut by a binder and shocked in the field to dry; eventually it is loaded onto wagons drawn by horses or mules, carried to the barn, and threshed in the barn by belt power provided by a stationary tractor. The straw is blown to the place where it will be stored until it is forked into a stall for bedding; the grain is blown into a bin. Homeostasis in the Amish community, on each farm and in each field, places purpose and mechanism in subordinate roles.

6. *It is impossible to reduce the higher level to the lower.* Since each level has its own characteristic structure and emergent qualities, reduction is

impossible without losing those qualities. There is no greater reality in the parts than in the whole; they are equally real. This concept is widely overlooked or ignored by most scientists today. In 1972, P. W. Anderson wrote a paper entitled, "More Is Different."[7] He emphasized that each level of organization is more than layers of atoms, that each level has its own laws every bit as fundamental as the fundamentals of physics. Nevertheless, we have been taught that many of the common sense observations all around us are an illusion and that the component parts are the reality. It is like saying that the liquidity of water is an illusion because hydrogen and oxygen are gases.

When we look at the major problems of modern agriculture—soil loss, chemical dependence, increasing dependence on fossil fuel, loss of the family farm, more corporate farms, an expansion of agribusiness—we see that most of them are the consequence of too much reductionism. It is understandable. We live in a society dominated by scientific reductionism. The problem is deep and probably resides in the history of science and technology. Both have been deeply influenced by physics, which we placed at the top of the hierarchy because we believed that once we understood the building blocks of nature, all else would be chemistry. Consequently, the most acceptable explanations in science have been in physical rather than biological terms. One can see where this leaves the biologists. They were simply discouraged from postulating scientific laws for the various biological levels because scientists in general regarded biological phenomena as too indeterminate for safe prediction.

This justification for and emphasis on reductionism in science and technology in general has been carried over from physics to biology to agricultural science. At the extreme, it appears that a license has been issued to an agricultural establishment of scientists and technologists to function as salesmen of industrial farm inputs. These salesmen of everything from fertilizer to computers push their products with an abundance of quantitative documentation of their performance. They define problems for farmers— problems farmers scarcely knew they had—and then sell them the remedies. On the slickest of paper, these salesmen may display impressive tables and graphs, products of highly mathematical econometric models developed and expanded by econometricians in the department of agricultural economics back at the land-grant university. In some respects, these academics are worse for the farmer than the salesmen, for they function as the agricultural priests who pass on the faith in reductionistic thinking. As Maurice Telleen, editor of *Draft Horse Journal*, says, "An unwary farmer is a quarry to be mined by men with minds with one dimension."

Meanwhile, many farmers, along with the rest of society, have come to distrust their common sense observations. They have been made to feel that they are an illusion. Fewer and fewer farmers think about their farms as an ecosystem. A good farmer will continue to look at a particular hillside and see what possibilities it offers in the total scheme of things, which includes his farm as a whole, its history, his family, and the aptitude of everyone in the family. An agricultural economist usually does not consider any of this because he, along with most of society, still distrusts many common sense observations.

We have also been warned away from analysis at higher levels of organization because, as we move up the hierarchy of the sciences from physics to biology, it appears that various attributes become so complex and variable that, at best, experts in systems analysis are necessary to keep track of it all. But as Eugene Odum has said, "It is an often overlooked fact that other attributes become less complex and less variable as we go from the small to the large unit." Odum reminds us that there are "homeostatic mechanisms, that is, checks and balances, forces and counterforces, [which] operate all along the line, and a certain amount of integration occurs as smaller units function within larger units." [8] The rate of photosynthesis of a forest community, in an example Odum provided, is less variable than that of individual leaves or trees within the community because when one part slows down, another may speed up to compensate. Odum went on to say that "when we consider the unique characteristics which develop at each level, there is no reason to suppose that any level is more difficult or any easier to study quantitatively."

As we move up the hierarchy, "more is different," partly because emergent qualities develop at each level. Although the findings in the study of one level may be useful in understanding the next level above, they never completely explain the next level. For example, with water all the attributes of the higher level (the liquid quality of water) were not predictable by knowing only the properties of the lower level (the gases of hydrogen and oxygen).

The ecosystem, as a volumetric "thing," can be a useful conceptual tool when thinking about sustainable agriculture. We can put mental boundaries around whatever part of the landscape we want to examine. A good farmer is constantly making mental cubes or spheres that include the vertical as well as the horizontal dimensions of a field or farm or even a farm community. To call them cubes or spheres oversimplifies what a good farmer does. He is at once an accountant or acknowledger of what passes through the various boundaries. If he simplifies his farm by selling off the livestock, for example, he has simplified more than his work. Mostly because of the nature of the economic system, he has simplified his thought, a kind of tragedy, for what farms need now, desperately so, is more thought by their owners. But just as important, perhaps, is being able to acknowledge that the ecosystem obeys the same laws as the other integrative levels. We can take an example from a lower level in the hierarchy and relate it at the ecosystem level and have a case that is stronger than an analogy, although perhaps weaker than an homology.

THE HIERARCHY OF DESCENT

As useful as the ecosystem concept may be for those of us thinking about sustainable agriculture, it is not enough. As suggested earlier, we cannot think about agriculture without thinking about *populations* or *species*— cows, hogs, sheep, chickens, wheat, corn, rice, etcetera. Just because these domestic species and their thousands of populations failed the test of "thinghood," their importance is not diminished. So we need to talk about species and populations that fall under the hierarchy of descent, a hierarchy that must have its own rules or laws. Although they are yet to be written, they must be independent of the laws for the hierarchy of structure.

It is useful and accurate to think of each of the numerous species operating in a prairie—be they song birds or grasses or bacteria—as *boundaries of biological information*. Prairie chickens don't mate with upland plovers or scissor-tailed flycatchers, let alone with big bluestem. The gene pools of the different species are mostly bounded. Plants representing 237 species have been counted on a square mile of native prairie in Nebraska.[9] When we add to those the species of vertebrates, algae, fungi, bacteria, actinomycetes, annelids, mollusks, centipedes, millipedes, insects, and arachnids, the quantity of biological information is awesome. As part of the hierarchy of descent, every species fits somewhere in a phylogenetic tree, as part of a phylogenetic hierarchy. Birds are more related to one another than they are to reptiles, and they are "higher" on the evolutionary scale because they came from reptiles, not the other way around. Anyone who has thought about it is struck by the range in size and shapes of all life forms from the viruses to the whales and redwoods.

There is another kind of diversity, the diversity within a species. Although our human viewpoint may be such that we may scarcely notice the variations within a population of meadowlarks on the prairie, meadowlarks do vary. So does a population of compass plants and so do populations of all the other species. Some are more noticeable and more variable than others. The variation in dogs, for example, is dramatic.

This range of variation within a species has something to do with the remarkable extremes we find in species dispersion. Humans, of course, are the most cosmopolitan of all and are highly variable. On the other hand, fellow creatures on this planet are noteworthy because they are restricted in their range; there is probably very little variation within their populations. For example, a fungus species called *labaulbenia* grows exclusively on the back part of the elytra of a beetle. The beetle is endemic to limestone caves in southern France. There are fly larvae that develop exclusively in seepages of crude oil in California. Larvae of a cerain fruit fly species develop only in the nephric grooves beneath the flaps of the third maxilliped of a land crab endemic to certain islands in the Caribbean.

It is not uncommon to discover that genetic variation is on the increase within a species whether or not that species is expanding its range—that is, the biological information within the boundary of that species is increasing. Two biologists, D. R. Brooks and E. O. Wiley, have postulated that at a certain point when biological information is accumulating within a species, when new genes are being added, the information of that species is becoming more chaotic, more entropic.[10] It seems to follow that the flow of energy through such a species also becomes more chaotic or disordered. Brooks and Wiley believe that, under certain circumstances, high information entropy actually causes populations to divide or speciate. This creation of two information boundaries, two species, where one once existed has some emergent qualities. First of all, whether Brooks and Wiley are right or not as to cause, it seems appropriate to assume that each of the two new populations has less information to contend with than does the parent population from which they were derived. (Even if we divide a herd of Holstein cows in half, there is less variation in each of the two resultant populations than in the original.)

With speciation, the ecosystem is usually a beneficiary in that its energy flow is more ordered, less chaotic; new niches are formed, new relationships are established. The ecosystem may become more stable. But even species diversity within an ecosystem has its limits. There can be too much. For years, natural preservationists and ecologists alike believed that more diversity within an ecosystem meant more stability. This is not necessarily true. R. M. May has written a book that summarizes the numerous examples showing that diversity and stability do not necessarily go together.[11] Many of his examples are of simple communities that were more stable than similar but more complex communities. In other words, information entropy can exist within the bounds of an ecosystem just as information entropy can exist within the bounds of a species. If Brooks and Wiley are right in their contention that information entropy within species is responsible for the evolution of new species, the possibility follows that information entropy at the ecosystem level is responsible for the evolution of new kinds of ecosystems.

It requires but little imagination of farming or gardening experience to realize that a particular farm can have too many species. If there are too many kinds of animals, there will be too many chores to do. There can be too many crops to tend. But in our time, too many species on a farm is seldom the problem; the usual problem is too few.

To think only about a high information agriculture, meaning lots of species diversity on the farm, is not enough. We have to think also about taking advantage of the natural integrities that exist in all the diversity. In a polyculture of plants, one species that fixes nitrogen complements others that are unable to fix nitrogen. Another species may do a better job of pulling up trace elements necessary for the nitrogen fixer and other species as well. This is diversity with complementarity; much, but not all, of it is fortuitous. On a diverse farm, we see complementarity. Some of the corn fed to cows passes through with the manure and is eaten by hogs and chickens. Both the cow and hog manure contain meals for chickens. With so many vitamins and minerals, bright orange yolks loaded with barnyard energy and nutrients fall into a skillet and contribute to the health and quality of life for the human. Here diversity of species slows entropy but complementarity is as important as diversity. It is important to remember, however, that without diversity we lack the basis of complementarity. Natural integrities are dependent upon diversity.

Up to now, we have considered two hierarchies in our attempt to discover the unifying concept for sustainable agriculture. The first is the hierarchy of objects that includes life forms. We have seen that there are certain common laws for the various integrative levels in this hierarchy, from the atom to the ecosystem. They are very general laws but useful, nevertheless. The second is the hierarchy of descent from the first cells to modern mammals, flowering plants, etcetera. This hierarchy of descent is the consequence of speciation, and with speciation, information entropy is reduced. Whether information entropy is responsible for speciation is still an open question among biologists. For our purposes, it is important to consider only that division of an information package reduces information entropy and probably energy entropy as well. In other words, for biological information to be used in an optimum manner, new boundaries have to be established because, at some point, a restriction of information is more necessary for stability than a free flow of information.

Agriculture comes out of nature, out of the two hierarchies described here: hierarchies that intertwine but obey certain laws independently. Our domestic plants and animals are more accurately thought of as relatives of wild things than as property of humans, even though most farmers and agricultural researchers think of them as property. To think about agriculture is to think about life, and to think about life is to think about the way nature has bounded its information to live on the earth and why it is so packaged. What are the rules? Are any of them universal up and down the hierarchy of descent? Are they as universal as the laws of integrative levels of structure? If we don't know the laws or rules, isn't it possible that some of our problems with agriculture are the direct result of this ignorance? There are a lot of "ifs" here, but there may also be an agenda for coming to grips with the problems of agriculture.

THREE FACTORS COMMON TO BOTH HIERARCHIES

One factor that appears to be of importance in a consideration of both hierarchies is scale. In the hierarchy of structure, scale increases upward from atom to organism. Each level has natural boundaries, so each level is unambiguous. At the ecosystem level this changes because humans draw the boundary lines. As scale increases both upward through the hierarchy or at only one level, whether it is a cell, an organism, a field, farm, or farm community, *a linear increase in size is seldom if ever attended by a linear increase in relationships.* An ostrich egg as a large cell necessarily has different properties from a hummingbird egg. For purposes of illustration, a better example is found among potatoes grown by the Peruvian Indians high in the Andes.[12] These potatoes are generally smaller than "improved" or commercial varieties, and estimates of the number of natural varieties range from "well over 400" to more than 2,000. In spite of what experts may think, to increase the size of the native potatoes could be a serious mistake. Larger potatoes would be harder to cook at the high elevations where the people farm and live and they would be more watery; additionally, since the skin contains so many nutrients, an increase in size would greatly change the surface to volume ratio, so that the potatoes would yield proportionally more starch than nutrients.

The size of the individual potatoes involves considerations formally understood by the laws or rules associated with the hierarchy of structure. The number of genetic varieties numbering from 400 to 2,000 involves considerations formally understood in our thinking about the hierarchy of descent. Scale is an important factor in both hierarchies.

Farmers assume the responsibility to manage an ecosystem, an enormous conceptual task partly because the boundary of the ecosystem, unlike the other levels in the hierarchy up through organisms, changes with scale. It is somewhat arbitrary where we draw the lines for a human-managed ecosystem, for it depends on what we wish to measure or to observe going in and out through the boundaries, whether it is a garden, field, pasture, farm, or farm community.

When speaking of ecosystems, we are talking about a community of organisms and the physical world that connects them. We are also talking about all the spaces between as components, not as void, because the actions between boundaries occur in those spaces. The ability of space to expand is another reason why scale considerations are inescapable.

Information is common and essential in both hierarchies, too. On the farm we may be talking about a milk cow as an organism or about a garden, patch, field, farm, or farm community as ecosystems and as we do, we are at home in the hierarchy of structure. On the same farm we may be talking about the dairy herd, the hogs, or chickens, all local populations of species that are relatives of wild things and fall under the hierarchy of descent. The information stored in nearly every cell of an organism is responsible for its growth, development, daily living, and reproduction. The information packaged in each somatic cell is part of a cloned population of such information packages; the local dairy herd population is part of a global breed, which is related to all the domestic breeds and ultimately to wild cattle.

While all of this biological information is subject to Darwinian natural selection, cultural information on the other hand, is subject to Lamarckian selection (in which environmental changes cause structural changes in animals and plants). Acquired cultural information can be handed down; although it is more vulnerable than biological information as a whole, it is less vulnerable than most of us think. I heard recently that even though Taoism has no official status in China, it is alive in each village even today. Even so, less cultural information is present on the earth than is biological information. To destroy biological information and expect cultural information to take its place on a one-to-one basis is risky. This is one reason the unsettling of America has been such a tragedy. Countless "bits" of cultural information have been lost in one generation.

The third factor common to both hierarchies is energy. Our natural world, assembled independently of human hands, is a world in which life, as the main or most interesting attraction, has had to rely almost exclusively on what we might call *contemporary sunlight*. The fact that sunlight is dispersed imposes certain limits on its use. The physical structures of the earth impose certain patterns of storage and transfer of energy. The various rocks and densities of air and water have their respective specific heats; a pound of air will hold about one-fifth as much heat as a pound of water. A pound of rock will hold a fifth to a fourth as much heat as a pound of water. Nature's information system accommodates all of these factors that influence the chemical interactions in the soil. It is staggering to contemplate, but this living world more than copes; it evolves and thrives. Except when an ecosystem uses fire, nature's biota operates at low temperatures.

By contrast, the industrial economy runs on energy that is not contemporary. It uses fossil fuels whose age must be measured in the tens of millions of years or the nuclear fuels that are as old as the universe itself. In the use of all these forms of energy, high temperature is the rule. The technologies that respond best to such energy more than invite us to expand the scale of a farming operation. This is how the industrial economy has elbowed its way into agriculture, into processes that are fundamentally biological in nature. Scale expands in nearly every phase of the operation. We use fossil fuels to control our competitors—the weeds, insects, and pathogens. We compact the earth with heavy equipment powered by fossil fuels. Fertilizer is processed at high temperatures mostly with natural gas as the feedstock. If nature had not been so resilient, we might have backed out long before such a complete infrastructure was put in place to exploit highly concentrated, ancient energy. We might have started earlier to accommodate a more sensible agriculture, one that contains a biological, physical, and chemical structure tuned to contemporary solar energy. All biological information, whether it is in microbes, higher vertebrates, or flowering plants, is tuned to a sun-powered earth. When humans destroy this information, they must substitute cultural information, apply more energy, or do both. Usually much more energy is applied than cultural information, and ecological capital is wasted or polluted. Soil erodes and is treated with chemicals for fertility and pest control.

THE INTERACTION OF SCALE, INFORMATION, AND ENERGY

The importance to agriculture for understanding the three interacting dynamics common to both hierarchies may be best clarified when we contrast two ecosystems: a modern cornfield and tall-grass prairie as an example of a natural ecosystem. If one could tease out the genetic instructions from a cell of a corn plant and type the instructions necessary to make the corn plant grow and prosper, the instructions would fill thousands of volumes, perhaps enough to fill a small house. If we could discover and record the biological information in all native prairie plants, it would fill all the libraries of the world and then some.

Biological information makes it possible for a native prairie ecosystem to remain intact without the human, provided all trophic levels that make up the prairie community are intact. Even where humans have turned native prairie into pasture for cattle by building fences and ponds, only a small amount of cultural information and intelligence is necessary because the biological information remains high. A ranch in the Flint Hills of eastern Kansas consists of nearly 6,400 acres of native prairie—nearly ten square miles. It has never been plowed and supports around 1,700 head of cattle during the grazing season. Aside from additional help needed for fencing, branding, castrating, loading, and unloading, that entire acreage is managed by a single cowboy who usually rides the range in a pickup truck. This is typical management practice for the region. Protecting that rolling country from erosion requires so little effort because the roots and leaves and small creatures of the soil of that diverse biota protect it and keep it stable. The roots are always there, as are the leaves, except when the prairie is burned. Biological information translates into chemical diversity and prevents epidemics of insects and pathogens, so that the cattle can have most of the grass. Ranchers never add fertilizer. They stock the land at the rate of about one cow per four acres. What prevents overstocking of that range land is the intelligence of the owner and the cultural information handed down. When signs of overstocking appear, the cattle are shipped out. The fossil energy input per acre in that prairie is practically nil. The ecosystem, powered by the sun's energy, provides its own fertility. Most of the weight of the cattle is water. The carbon and nitrogen ultimately come from the air. When the cattle are hauled off, the nutrients that came from the soil must be replaced by solar-powered root pumps extending into the subsoil or the rock a few inches to a few feet below the surface. They are the same mines that supported the American bison.

When we are talking about ecosystems being directed to meet the human purpose, the scale of operation becomes critical. When the biological information per acre is high, the cultural information being applied per acre can be very low. This means that the ratio of acres to humans can be very high as it is with our Flint Hills ranch. There are products of the industrial revolution on this prairie—the barbed wire, the steel fence posts, the small dams built with bulldozers, the pickup truck, even the breeding that has produced some of the highly efficient weight-gaining cattle. But overall, nature dominates this landscape. The fossil energy input per pound of protein, as well as the fossil energy input per acre, is very low. In this system, the ratio of biological information to cultural information per pound of produced protein may be the highest of any place on the land surface of the earth. As a food-producing ecosystem for humans, it is highly resilient, although much of the mammalian diversity is now gone.

On September 12, 1806, Zebulon Pike stood on a high hill just a few miles from this ranch and saw the prairie wolves, panthers, American bison, wapiti (elk), deer, and antelope. In less than one hundred years, white settlers eliminated their competitors in the food chain—the wolf, the panther, and the Indian—and the wild ungulate herbivores—the bison, wapiti, deer, and antelope—in favor of their own domestic ungulate, beef. In spite of these extirpations, the ranch ecosystem has remained essentially intact because humans and cattle are sufficiently close as ecological analogs to the larger wild grazers and their predators. Every lost and substituted creature was a mammal. Most of the prairie vegetation changed but little. But one important lesson remains: even in this instance, before ranchers could alter that wild ecosystem for their use, they had to simplify it.

Now let us move our attention to 6,400 acres of former prairie land in an Amish settlement in northern Indiana. With an average of 80 acres per Amish family, a diversified agriculture here would support 80 families. Even if the livestocking rate were four times as high, as I suspect it could have been, I doubt that management would require even four times the effort, time, and thought that our Flint Hills ranch requires. The point is, however, that if the soil and water are to be maintained where till agriculture is practiced, the jobs done by the prairie biota now must be done by humans and their domestic species. With the loss of biological diversity, the price for sustainability must be paid from elsewhere. The Amish substitute cultural information for biological information—muscle power from the draft animals and from humans. The Amish use small engines on some pieces of equipment, such as balers, which are pulled by draft animals. They also use tractors for belt power to run stationary equipment such as threshing machines. Consequently, the Amish may use more fossil fuel per acre of production than the Flint Hills ranch uses. Of course, they produce more food per acre, but they may not necessarily use fewer fossil-fuel calories to produce 1,000 calories of food. Accurate comparisons are complicated because what the human harvests from the rangeland is mostly red meat, whereas the Amish farm produces grain and other plant products, draft power, milk, as well as meat. But for the Amish to farm the 6,400 acres, they must increase the ratio of people to acreage 150 to 200 fold. One reason for this increase is that with till agriculture, more attention must be paid to the vertical dimension, particularly below the surface.

The same northern Indiana acreage farmed with conventional equipment by industrial farmers would likely support fewer than half the number of families and probably one-third the number of people. (Amish families are larger than average.) The species diversity on the Amish farm is much higher than on the conventional farm—the Amish usually grow more crops and keep more kinds of animals. A larger measure of their fertility comes from crop rotation and manure. Many Amish farms do use commercial fertilizer and pesticides, but the extent of their use is usually much more restricted than on the average conventional farm.

In western Kansas, two related farm families may own and farm as many as 6,400 acres. Because this must be a dryland farming operation, they will probably devote most of their acreage to growing sorghum and wheat. There may be two generations associated with such an operation, or the families or two siblings may be farming together; there are many combinations. With seasonal outside help, this operation is viable for the present. Energy takes the place of information, and as long as the liquid fuel is available, farming this flat country can continue. Erosion is not a serious problem.

But what if such a 6,400-acre western Kansas farm were to use only the solar energy captured on that farm for field work, growth, and harvest? With a ratio of one draft animal to sixteen acres or more, the most energy-efficient way of doing the work would be to use the muscle power of draft animals and humans; with this ratio, if some of the grain or stover were turned into alcohol for tractor fuel, more energy would be required. Of course, if tractors were used, fewer people would be necessary. If all 6,400 acres were planted to short-grass prairie, it could still support livestock and be cared for in the same manner as our 6,400-acre Flint Hills ranch. Eighty acres wouldn't support even an Amish family in this area beginning around the 100th meridian (western Kansas), where evaporation is in excess of rainfall. Without irrigation, the portion of each year's crop that would have to go into the draft animals to grow the crop would be so great that the harvest available to sell could not be counted on to support the most frugal Amish family. Although a few Amish remain in Kansas, many more have moved. The Amish possess cultural information for farming in humid climates. Of those who have stayed, most farm with tractors.

But there is an important consideration raised by this comparison of the economies of northern Indiana and western Kansas. Even in Iowa, where there is an abundance of rainfall, a farm powered by draft animals will require more acres, although less energy, to support the animals than will be necessary to support tractors. Draft animals require a diet while tractors mostly require energy. To grow a balanced diet requires that we feature diversity, which forces crop rotations. In many cases, certain land now regarded as marginal or difficult can become permanent pasture. This may be land along a stream which, on an industrial farm, is now "farmed right through." This kind of accounting forces us to realize that if we are to practice till agriculture without ruining the land, our standards have to be biological and not industrial. When the expectations of the land can't be met with conventional crops, the land should be returned to nature or to a vegetative structure which resembles that of nature.

SUMMARY

The ecosystem is a category immediately above organism in the hierarchy of objects from atoms through cells, tissues, and so on up. All the levels in the hierarchy obey the same laws of integration. Because these laws are common for every level, examples drawn from a lower level and applied to the ecosystem are of greater utility as an explanation than if they were mere analogies.

The hierarchy of descent that we observe in the earth's biota is the consequence of speciation. Species can be viewed as information boundaries that, once established, were less information and energy entropic than their immediate ancestral species. The laws or rules governing this hierarchy have not been formulated.

Laws or rules govern the two hierarchies that are independent of one another, but the two are intricately intertwined and probably have been over the course of evolution. Common to these two intertwining hierarchies of objects and descent are three interacting dynamics: those of scale, information, and energy. Scale can expand or decrease, and as it does, the relationship between information and energy changes dramatically. For agriculturists, information is of two types: biological and cultural. Biological information is subject to Darwinian selection and is more reliable than cultural information, which is Lamarckian in nature, but both types of information are needed. There can be too much biological information in an ecosystem and on a farm. Too much diversity can spell instability, although this is seldom a problem on the farm. Most farms suffer from too much energy, too large a scale, and too little information. An abundance of energy, especially of the fossil-fuel variety, has a way of homogenizing the farm ecosystem. With too much energy, there is a strong incentive to reduce diversity in order to simplify the farm structure and, in so doing, to break down the complementarities in the biotic and cultural diversity on the farm.

Discovering the right balance of cultural and biological information and the balance between information and energy, given the scale of an operation, is necessary for sustainability, a synonym for homeostasis. It is out of an understanding of these dynamics that a taxonomy can begin to develop, a taxonomy that can comfortably include everything from a small garden to a large ranch.

We may finally come to understand all of this as a kind of harmony. In this respect, Wendell Berry has already summarized what I have said here: [13]

In a society addicted to facts and figures, anyone trying to speak for agricultural *harmony* is inviting trouble. The first trouble is in trying to say what harmony is. It cannot be reduced to facts and figures—though the lack of it can. It is not very visibly a function. Perhaps we can only say what it may be like. It may, for instance, be like sympathetic vibration: "The A string of a violin . . . is designed to vibrate most readily at about 440 vibrations per second: the note A. If that same note is played loudly not on the violin but near it, the violin A string may hum in sympathy." This may have a practical exemplification in the craft of the mud daubers which, as they trowel mud into their nest walls, hum to it, or at it, communicating a vibration that makes it easier to work, thus mastering their material by a kind of song. Perhaps the hum of the mud dauber only activates that anciently perceived likeness between all creatures and the earth of which they are made. For as common wisdom holds, like *speaks to* like.

Notes

NOTES TO CHAPTER THREE

1. U.S. Department of Agriculture, *A Time To Choose: Summary Report on the Structure of Agriculture* (Washington, D.C.: U.S. Department of Agriculture, January 1981), pp.23–27.
2. This public attitude, which has shown up in Gallup and Harris polls, is supported by social researchers studying the impact of large-scale farms on community development and democratic politics. See Isao Fujimoto, "Farming in the Rockies: A Humanistic Perspective," in *The Future of Agriculture in the Rocky Mountains*, ed. E. Richard Hart (Salt Lake City, Utah: Westwater Press, 1980), pp. 74–75; and Walter Goldschmidt, *As You Sow: Three Studies in the Social Consequences of Agribusiness* (Montclair, N.J.: Allanheld, Osmun, 1978).
3. Kenneth L. Robinson, "The Impact of Government Price and Income Programs on Income Distribution in Agriculture," *Journal of Farm Economics* 47, no. 5 (December 1965): 1225–1234.
4. In 1979, there were 6.2 million persons living on farms, or 3 percent of the national population. That number is based on the current definition of a farm, which does not count places with less than $1,000 a year in farm sales. That is also a way of declaring, for the purposes of the government, that noncommercial agriculture does not exist. USDA, *A Time to Choose*, p. 34.

5. In 1900, there were 838,591,744 acres of American land in farms; by 1978, the total was 1,029,694,535 acres. Most of the increase in farmland occurred before World War II, but most of the decline in the number of farms took place, rather precipitously, after the war. See U.S. Bureau of the Census, *Thirteenth Census of the United States*, vol. 5, *Agriculture* (Washington, D.C.: U.S. Government Printing Office, 1913), p. 28; *1978 Census of Agriculture* (Washington, D.C.: U.S. Government Printing Office, 1981), vol. 51, pt. 51, p. 1.

6. See James Turner, *The Chemical Feast* (New York: Grossman, 1970), chap. 5; and Jim Hightower, *Eat Your Heart Out* (New York: Crown, 1975).

7. USDA, *A Time to Choose*, p. 75.

8. Peter Manniche, *Living Democracy in Denmark* (Toronto: Ryerson Press, 1952), p. 236.

9. James Risser, "A Renewed Threat of Soil Erosion: It's Worse Than the Dust Bowl," *Smithsonian* 11, no. 12 (March 1981): 127.

10. The ecological approach to farming appears to have more serious support in some European countries than in the United States. See, for instance, the officially sponsored Dutch report, R. Boerringa, ed., *Alternative Methods of Agriculture* (Amsterdam: Elsevier, 1980); and Gil Friend, "Biological Agriculture in Europe," *CoEvolution Quarterly*, Spring 1978, pp. 60–64. In the United States, research and support have so far come from outside government and established agricultural schools. See, for example, Wes Jackson's *New Roots for Agriculture* (San Francisco: Friends of the Earth, 1980); and Richard Merrill, ed., *Radical Agriculture* (New York: New York University Press, 1976), especially pt. 5.

11. The book was first published in 1927 by Harper & Brothers. Rölvaag was a professor of Norwegian literature in Minnesota and wrote in his native Norse. Two editions are available at this writing: the 1927 edition is still published by Harper & Row, in hardback, and there is a 1965 edition, also from Harper & Row, in paperback.

NOTES TO CHAPTER FOUR

1. A. B. Hulbert, *Soil: Its Influence on the History of the United States* (New York: Russell and Russell, 1969).

2. Hans Jenny, *The Soil Resource, Origin and Behavior* (New York: Springer-Verlag, 1980).

3. Jenny, *The Soil Resource*.

4. M.A. Walters, "Soil Carbon and Nitrogen Trends with Varied Land Use Histories, 1930–1980" (M.S. thesis, University of Missouri, 1981).

5. H.H. Bennett, *Soil Conservation* (New York: McGraw-Hill, 1939).

6. Bennett, *Soil Conservation*.

7. D.M. Kral, ed., *Determinants of Soil Loss Tolerance*, special publication no. 45 (Madison, Wis.: American Society of Agronomy, 1982).

8. G. Friend and D. Maher, *Biofuels Development and Soil Productivity* (Sacramento, Calif.: Office of Appropriate Technology, 1982).

9. N.A. Berg, "Resources Conservation Act: New Window on the Future," in *Economics, Ethics, Ecology*, ed. W.E. Jeske (Ankeny, Iowa: Soil Conservation Society of America, 1981), pp. 134–141.

NOTES TO CHAPTER FIVE

1. Actually, Taoist writers dealt with flood control more than irrigation, but their water philosophy of noninterference has many modern applications. For a useful discussion of their ideas, see Joseph Needham, *Science and Civilization in China* (Cambridge: Cambridge University Press, 1971), vol. 4, pt. 3, pp. 235–251.

2. Aldo Leopold, *Round River* (New York: Oxford University Press, 1953).

3. Robert Curry, "Watershed Form and Process: The Elegant Balance," *Co-Evolution Quarterly*, Winter 1976–1977, pp. 14–21.

4. Much of what follows is based on these sources: Gamal Hamdan, "Evolution of Irrigation Agriculture in Egypt," in *A History of Land Use in Arid Regions*, ed. L. Dudley Stamp (Paris: UNESCO, 1961), pp. 119–142; Karl Butzer, *Early Hydraulic Civilization in Egypt* (Chicago: University of Chicago Press, 1976); Desmond Hammerton, "The Nile River—A Case History," in *River Ecology and Man*, eds. Ray Oglesby, Clarence Carlson, and James McCann (New York: Academic Press, 1972), pp. 171–214; John Waterbury, *Hydropolitics of the Nile Valley* (Syracuse, N.Y.: Syracuse University Press, 1979); and Robert Tignor, "British Agricultural and Hydraulic Policy in Egypt, 1882–1892," *Agricultural History* 37, no. 2 (April 1963): 63–74.

5. Karl Wittfogel, *Oriental Despotism: A Comparative Study of Total Power* (New Haven, Conn.: Yale University Press, 1957). This is a controversial effort to link water technology to centralized authority in Asian history. Consult also, among the large number of titles on this theme, Marvin Harris, *Cannibals and Kings: The Origins of Cultures* (New York: Random House, 1977), chap. 13; and William Mitchell, "The Hydraulic Hypothesis: A Reappraisal," *Current Anthropology* 14 (December 1973): 532–534.

6. Bureau of Reclamation, *Water and Land Resource Accomplishments: 1977* (n.p., n.d.), p. 1.

7. Philip Fradkin, *A River No More: The Colorado River and the West* (New York: Knopf, 1981), p. 16.

8. I have discussed this process in my "Hydraulic Society in California: An Ecological Interpretation," *Agricultural History* 56, no. 3 (July 1982): 503–512.

9. U.S. Water Resources Council, *The Nation's Water Resources, 1975–2000* (Washington, D.C.: U.S. Government Printing Office, 1978), vol. 1, pp. 28–29. The 1975 coterminous U.S. rate was estimated at 159 million gallons a day withdrawn for agriculture out of a total of 339 million gallons a day.

10. C. Richard Murray and E. Bodette Reeves, *Estimated Use of Water in the United States in 1970*, U.S. Geological Survey Circular 676 (Washington, D.C.: U.S. Government Printing Office, 1972), pp. 22–23. Of the 140 million acre-feet of water withdrawn nationally for irrigation in that year, 82 million

acre-feet were considered "consumed"—that is, no longer available for stream flow or other uses.

11. "The Browning of America," *Newsweek* 97 (23 February 1981): 29–30. For a comparable decline, see David Todd, "Groundwater Utilization," in *California Water: A Study in Resource Management*, ed. David Seckler (Berkeley: University of California Press, 1971), pp. 174–189.

12. Curry, "Watershed Form and Process," pp. 17–18.

13. The state-financed California State Water Project, which takes the Sierra snowmelt from the north to agribusiness operations in Kern County and to the Los Angeles area, has a huge energy deficit. Pumping alone requires thirteen billion kilowatt hours a year, while the project generates only five billion. These figures are taken from Jeffrey Lee, "The Energy Costs of the California Water Project" (Pamphlet, Water Resources Center Archives, University of California at Berkeley, 1973).

14. Marty Bender and Wes Jackson, "American Food: S/Oil and Water," *The Land Report*, Winter 1981, p. 13. See also Margaret Lounsbury, Sandra Hebenstreit, and R. Stephen Berry, *Resource Analysis: Water and Energy as Linked Resources*, Water Resources Center, Research Report 134 (Urbana: University of Illinois, July 1978) pp. 118–175.

15. The salinity problem is discussed in the following: Myron Holburt and Vernon Valantine, "Present and Future Salinity of Colorado River," *Journal of the Hydraulic Division, Proceedings of American Society of Civil Engineers* 98 (March 1972): 503–520; Gaylord Skogerboe, "Agricultural Impact on Water Quality in Western Environment," in *Environmental Impact on Rivers*, ed. Hsieh Wen Shen (Fort Collins, Colo.: Privately published, 1973), pp. 12–1 to 12–25; George Cox and Michael Atkins, *Agricultural Ecology* (New York: W. H. Freeman, 1979), pp. 300–308.

16. J. F. Poland and G. H. Davis report that at least 30 percent of California's pumped land has subsided; in some places the drop is nearly thirty feet. "Land Subsidence Due to Withdrawal of Fluids," *Reviews in Engineering Geology* 2 (1969): 187–269.

17. Charles Goldman, "Biological Implications of Reduced Freshwater Flows on the San Francisco-Delta System," in Seckler, *California Water*, pp. 109–124; Robert Hagan and Edwin Roberts, "Ecological Impacts of Water Projects in California," *Journal of the Irrigation and Drainage Division: Proceedings of American Society of Civil Engineers* 91 (March 1972): 25–48. The latter is a most comprehensive account and a model of wishy-washy conclusions.

18. The danger posed by deteriorating dams is far more serious than the public realizes. One study indicates that dams are 10,000 times more likely to cause a major disaster than are nuclear power plants. Gaylord Shaw, "The Nationwide Search for Dams in Danger," *Smithsonian* 9 (April 1978): 36.

19. See, for example, the remarks by Joseph Sibley, Congressman from Pennsylvania, in *Congressional Record*, 21 January 1902, p. 836; and of Gilbert Tucker, editor of *Country Gentleman*, in *Congressional Record*, 13 June 1902, pp. 6723–6724.

20. Charles Howe and William Easter, *Interbasin Transfer of Water: Economic Issues and Impacts* (Baltimore, Md.: Johns Hopkins University Press, 1971), pp. 138–140, 144–145, 167. See also Richard Berkman and W. Kip Vicusi, *Damming the West* (New York: Grossman, 1973), pp. 17–23.

21. James Risser, "A Renewed Threat of Soil Erosion: It's Worse Than the Dust Bowl," *Smithsonian* 11, no. 12 (March 1981): 127. Other areas with serious erosion are Iowa, which has lost half of its splendid topsoil in a century, and the Palouse country of eastern Washington, losing seventeen million tons from a million acres of cropland—as much as 200 tons per acre.

22. Some engineers are beginning to question the old hubris. Elmo Huffman of California's Department of Water Resources writes: "We must learn that meddling is not synonymous with management. . . . In many cases, the wisest management is simply to preserve things as they now are, or, at the most, only attempt to heal the wounds so carelessly inflicted by man in the past." "Role of the Civil Engineer in Total Watershed Development and Management," in *Development of the Total Watershed*, Irrigation and Drainage Conference, American Society of Civil Engineers (1966), pp. 43–44. See also Gilbert White, "A Perspective of River Basin Development," *Law and Contemporary Problems* 22 (Spring 1977): 157–187, which suggests some changing attitudes.

23. A similar recommendation comes from the study team headed by the impeccably conservative agricultural economist Earl Heady. See National Water Commission, *Water Policies for the Future* (Washington, D.C.: U.S. Government Printing Office, 1973), p. 141. The Heady group, however, simply wanted no new projects authorized.

24. Bureau of Reclamation, *Water and Land Resource Accomplishments: 1977*, p. 8.

25. The Third World, facing far more intense population pressures, may be forced into more and more environmental violence to survive. Simply to keep pace with their food needs, they will require by the year 2000 some twenty-two million hectares of new irrigation land, according to the prognosis of the Food and Agriculture Organization in *Water for Agriculture*, E/Conference, 70/11, 29 January 1977, 4.

26. Leopold's "The Land Ethic" essay appears in his *Sand County Almanac* (New York: Oxford University Press, 1949). His son Luna Leopold is one of the foremost hydrological experts in the United States and author of such technically sophisticated works as *Water in Environmental Planning*. But nowhere does the son seem impressed with the father's ethical concern or seek to extend that concern to the human use of rivers. Once again, a gain in scientific credentials apparently has been achieved at the cost of moral vision.

NOTES TO CHAPTER SIX

1. For the energy invested per calorie of food, see George Cox and Michael Atkins, *Agricultural Ecology* (New York: W. H. Freeman, 1979), p. 618. For the energy to feed the rural populations of developing nations, see Walter

Vergara, "Energy Use in Processing," in *Changing Energy Use Futures*, ed. Rocco A. Fazzolare and Craig B. Smith (New York: Pergamon Press, 1979), vol. 3, pp. 1862–1870.

2. For the loss of one-third of our cropland topsoil, see David Pimentel et al., "Land Degradation: Effects on Food and Energy Resources," *Science* 194, no. 4261 (1976): 149–155. For soil loss studies within the past decade, see T.R. Hargrove, *Iowa Agricultural Home Economic Experiment Station*, Special Report no. 69, 1972; and U.S. General Accounting Office, *To Protect Tomorrow's Food Supply, Soil Conservation Needs Priority Attention*, CED-77–30 (Washington, D.C.: U.S. Government Printing Office, 1977).

3. For the figures on irrigation and overdraft, see U.S. Water Resources Council, *The Nation's Water Resources: 1975–2000* (Washington, D.C.: U.S. Government Printing Office, 1978), p. 18.

4. Pimentel et al., "Land Degradation," p. 152.

5. Robert den Bosch, *The Pesticide Conspiracy* (New York: Anchor Press/ Doubleday, 1980), p. 24.

6. R.A. Friedrich, *Energy Conservation for American Agriculture* (Cambridge, Mass.: Ballinger, 1978), p. 55.

7. Wes Jackson, *New Roots for Agriculture* (San Francisco: Friends of the Earth, 1980), pp. 26, 28.

8. Harold F. Breimyer, "Outreach Programs of the Land Grant Universities: Which Publics Should They Serve?" Keynote address at Kansas State University for conference with same name and title, 1976.

9. For farm population and number of farms, see U.S. Department of Agriculture, *Agricultural Statistics* (Washington, D.C.: U.S. Government Printing Office, 1977), pp. 424, 434. For the population of horses and mules, see Gerald W. Thomas et al., *Food and Fiber for a Changing World: Third Century Challenge to American Agriculture* (Danville, Ill.: Interstate Printers and Publishers, Inc., 1976), p. 105.

10. U.S. Bureau of the Census, *Statistical Abstracts of the U.S.* (Washington, D.C.: U.S. Government Printing Office, 1982–1983), pp. 662–663; and personal communication with Allen Smith of the USDA Economic Research Service in Washington, D.C.

11. For the percent of food that is processed, see John D. Buffington and Jerrold H. Zar, "Realistic and Unrealistic Energy Conservation Potential in Agriculture," in *Agriculture and Energy*, ed. William Lockeretz (New York: Academic Press, 1977), pp. 695–711. For the Standard Industrial Classification rankings, see John S. Steinhart and Carol E. Steinhart, "Energy Use in the U.S. Food System," *Science* 184, no. 4134 (1974): 33.

12. For the energy sources for the food-processing industries, see Samuel G. Unger, "Energy Utilization in the Leading Energy-Consuming Food Processing Industries," *Food Technology* 29, no. 12 (1975): 33–45. For energy input in vegetable canneries, see Vergara, "Energy Use in Processing," p. 1864.

13. For the solid wastes, see U.S. Bureau of the Census, *Statistical Abstracts*, p. 206. For the wastewater, see "Energy Use in Processing," p. 1864.

14. See note 10, and for the estimated $21.2 billion in government farm price supports, see Susan Tifft, "Farmers Are Taking Their PIK," *Time* (25 July 1983): 14.

15. Rob Aiken, "Conservation in the Corn Belt," *Soft Energy Notes* 3, no. 6 (1981): 12–13; and personal communication with Ron Krupicka of the Center for Rural Affairs.

16. Friedrich, *Energy Conservation*, pp. 56, 69.

17. Buffington and Zar, "Realistic and Unrealistic Energy Conservation," pp. 700–701.

18. For 100 million acres, see Peter C. Myers, "Why Conservation Tillage?," *Journal of Soil and Water Conservation* 38, no. 3 (1983): 136. For 10.5 million acres, see Arnold D. King, "Progress in No-Till," *Journal of Soil and Water Conservation* 38, no. 3 (1983): 160–161. For 85 percent and 153 million acres, see Maureen K. Hinkle, "Problems with Conservation Tillage," *Journal of Soil and Water Conservation* 38, no. 3 (1983): 201–206.

19. For the range of 30 to 50 percent, see Hinkle, "Problems with Conservation Tillage," p. 201; and Elyse Axell, "The Toll of No-Till," *Soft Energy Notes* 3, no. 6 (1981): 14–16. For extra nitrogen, see Axell, p. 15; and William Lockeretz, "Energy Implications of Conservation Tillage," *Journal of Soil and Water Conservation* 38, no. 3 (1983): 207–211. For net energy savings, see Dan Locker, "Problems Remain with No-Till Technology," *Prairie Sentinel* 1, no. 3 (1982): 8–10; Lockeretz, p. 211; and Axell, p. 14.

20. For metabolic pathways, see Hinkle, "Problems with Conservation Tillage," p. 203. For atrazine problems, see Hinkle, p. 204; and Axell, "The Toll of No-Till," p. 15.

21. For farmers who use minimum-till without herbicides, see Richard Thompson, "No-Till Beans Without Herbicides," *Prairie Sentinel* 1, no. 3 (1982): 8–10; and Dan Looker and Dennis Demmel, "A Low-Cost Minimum-till Program," *Prairie Sentinel* 1, no. 3 (1982): 11.

22. Wes Jackson and Marty Bender, "Saving Energy and Soil," *Soft Energy Notes* 3, no. 6 (1981): 7.

23. Amory B. Lovins, *Soft Energy Paths* (Cambridge, Mass.: Ballinger, 1977), p. 80.

24. A one-foot depth of soil weighs 2,000 tons per acre. See R. Neil Sampson, *Farmland or Wasteland: A Time to Choose* (Emmaus, Pa.: Rodale Press, 1981), p. 131. For energy in irrigated corn, see David Pimentel and David Burgess, "Energy Inputs into Corn Production," in *Handbook of Energy Utilization in Agriculture*, ed. David Pimentel (Boca Raton, Fla.: CRC Press, Inc., 1980), pp. 67–84.

25. For the references for all car information in this paragraph, see Amory B. Lovins and L. Hunter Lovins, *Energy/War: Breaking the Nuclear Link* (San Francisco: Friends of the Earth, 1980), pp. 93–94.

26. For retooling Detroit, see Lovins and Lovins, *Energy/War*, pp. 94–95.

27. Amory B. Lovins and L. Hunter Lovins, *Brittle Power: Energy Strategy for National Security* (Andover, Mass.: Brick House, 1982), p. 335.

28. For facts on milk production, see P.A. Oltenacu and M.S. Allen, "Resource-Cultural Energy Requirements of the Dairy Production," in *Handbook of Energy Utilization*, ed. Pimentel, pp. 364–378. For the number of gallons of fuel consumed annually by American cars, see U.S. Bureau of the Census, *Statistical Abstracts*, p. 619.

29. Jim Harding, "Ethanol's Balance Sheet," in *Tools for the Soft Path*, eds. Jim Harding et al. (San Francisco: Friends of the Earth, 1982), pp. 102–104.

30. J.O.B. Carioca and H.L. Arora, "Decentralized Integrated Systems for Biomass Production and Its Energy/Nonenergy Utilization," in *Improving World Energy Production & Productivity*, eds. L. Clinard, M. English, and R. Rohm (Cambridge, Mass.: Ballinger, 1982), pp. 161–210.

31. For the study of *Euphorbia*, see Charles Drucker, "Fuel Farms," in *Tools for the Soft Path*, ed. Harding et al., pp. 91–93.

NOTES TO CHAPTER SEVEN

1. The capacity and characteristics of the ethanol still were obtained from personal communication with Dr. Mike Ladisch of the agricultural engineering department of Purdue University at West Lafayette, Indiana.

2. For the Amish doubling the size of their population in the past twenty-five to thirty years, see Wendell Berry, "People, Land, and Community," in *Standing by Words* (San Francisco: North Point Press, 1983).

3. For examples of economic analyses and Robert E. Ankli's conclusion, see a symposium volume with Robert E. Ankli, "Horses Versus Tractors on the Corn Belt," *Agricultural History* 54, no. 1 (1980): 134–148.

4. W.A. Johnson, V. Stoltzfus, and P. Craumer, "Energy Conservation in Amish Agriculture," *Science* 198, no. 4315 (1977): 373–378. For an example of a sociological study of the Amish, see Victor Stoltzfus, "Amish Agriculture: Adaptive Strategies for Economic Survival of Community Life" in *Change in Rural America*, eds. Richard D. Rodefeld et al. (St. Louis, Mo.: C.V. Mosby, 1978), pp. 450–456.

5. For examples of farmers who have written on their observations of farms, see Gene Logsdon in this book; and Wendell Berry, "Restoring My Hillside Farm," *Country Journal* 10, no. 4 (1983): 45–49. For other observations on tractors, see note 2.

6. The 75–25 percent split in harvest energy was obtained by personal communication with professor Mark Schrock of the agricultural engineering department of Kansas State University, Manhattan, and by discussions with John Deere Company.

7. Personal communication with John Deere farm machinery dealer in Salina, Kansas.

8. John Hostetler, *Amish Society*, 3rd ed. (Baltimore, Md.: Johns Hopkins University Press, 1980), p. 123.

9. For the 1920 draft animal population in the United States, see U.S. Depart-

ment of Commerce, Bureau of the Census, *Historical Statistics of the United States: Colonial Times to 1970* pt. 1, series 570 and 572 (Washington, D.C.: U.S. Government Printing Office, 1975). For the acreage of harvested cropland, see U.S. Department of Agriculture, Economic Research Service, *Changes in Farm Production and Efficiency*, Statistical Bulletin no. 581 (Washington, D.C.: U.S. Government Printing Office, 1977), table 12. For the number of tractors, see table 30 in the same publication. For the operations that a tractor could not perform in 1920, see Ankli, "Horses Versus Tractors," p. 137.

10. Ankli, "Horses Versus Tractors," p. 145.

11. Hostetler, *Amish Society*, p. 126; and Johnson, Stoltzfus, and Craumer, "Energy Conservation," p. 376.

12. The energy data for field operations are from U.S. Department of Agriculture, Economic Research Service, *Energy and U.S. Agriculture: 1974 Data Base* (Washington, D.C.: U.S. Government Printing Office, 1976), vol. 2, pp. 26, 46, 86, 116.

13. The energy values for various feeds and dried distiller's grains are obtained from National Research Council, Subcommittee on Dairy Cattle Nutrition, *Nutrient Requirements of Dairy Cattle* 5th rev. ed. (Washington, D.C.: National Academy of Sciences, 1973) table 4, corn in line 95, p. 40; oats in line 169, p. 44; alfalfa in line 8, p. 38; dried distiller's grains in line 88, p. 40.

14. The weights of the tractors were obtained from personal communication with John Deere Company. For the 38,000 BTUs and 6 percent for repair, see David Pimentel and Elinor Terhune, "Energy and Food," *Annual Review of Energy* 2 (1977): 171–195.

15. For the 2,800 BTUs per gallon of ethanol, see R.S. Chambers et al., "Gasohol: Does It or Doesn't It Produce Positive Net Energy?," *Science* 198 (16 November 1979): 789–795.

16. For the 30,000 BTUs, see note 1.

17. M.E. Ensminger, Agricultural Research Service, U.S. Department of Agriculture, *Breeding and Raising Horses*, Agriculture Handbook no. 394 (Washington, D.C.: U.S. Government Printing Office, 1972), p. 37. The estimate of 700 hours of work per year was obtained by personal communication with Maurice Telleen, publisher of *The Draft Horse Journal*, Waverly, Iowa.

18. Wes Jackson and Marty Bender, "Horses or Horsepower?," *Soft Energy Notes* 5, no. 3 (1982): 70–73, 87.

19. John H. Martin, Warren H. Leonard, and David L. Stamp, *Principles of Field Crop Production* (New York: Macmillan, 1976), p. 159; and George A. Whetstone, Harry W. Parker, and Dan M. Wells, *Study of Current and Proposed Practices in Animal Waste Management* (Washington, D.C.: Office of Air and Water Programs, Environmental Protection Agency 430/9–74–003, 1974), p. 50.

20. For the number of farms and farm population in 1920, see U.S. Department of Agriculture, *Agricultural Statistics* (Washington, D.C.: U.S. Government Printing Office, 1977), pp. 424, 434. For the farm population and number of

farms at present, see U.S. Department of Agriculture, *Agricultural Statistics* (Washington, D.C.: U.S. Government Printing Office, 1981), p. 415. For the harvested cropland acreage for various categories, see table 12 in USDA, *Changes in Farm Production*, in note 9, and later discussion in this chapter. For the number of tractors, see table 30 in the same publication. For the per-acre productivity of cropland, see table 13 in USDA, *Changes in Farm Production*, in note 9. For the horse and mule population in 1920, see the U.S. Department of Commerce, *Historical Statistics*, in note 9. For the eight million horses and mules now in the United States, see U.S. Department of Agriculture, *Resources Conservation Act: Appraisal 1980*, Review Draft (Washington, D.C.: U.S. Government Printing Office, 1980), pt. 1, pp. 6–99.

21. For the statement on U.S. reserves of petroleum and natural gas, see William Splinter, "Energy Requirements in Managing for Food and Fiber Production" in *Renewable Resource Management for Forestry and Agriculture*, eds. James S. Bethel and Martin A. Massengale (Seattle: University of Washington Press, 1978), p. 67. For the statements on world reserves of petroleum and natural gas, see Maryla Webb and Judith Jacobsen, *U.S. Carrying Capacity: An Introduction* (Washington, D.C.: Carrying Capacity, Inc., 1982), pp. 43–45.

22. The 234 million was obtained from an Associated Press release in early May 1983.

23. The figures for total harvested cropland and export acreage were obtained from personal communication with Charles Cobb of the Economic Indicator and Statistics Branch in the Economic Research Service division of U.S. Department of Agriculture in Washington, D.C., in September 1983. For the one-fourth and three-fourths split, see Dan Luten, "What America Grows," *Landscape* 25, no. 1 (1981): 18–19.

24. For the million metric tons, see U.S. Department of Agriculture, Economic Research Service, *Feed: Outlook & Situation* (Washington, D.C.: U.S. Government Printing Office, November 1982), table 16, pp. 26–27. For the calculations by Dan Luten, see note 23. For the 10.9 ounces of meat per capita daily, see a United Press International release in *The Salina Journal* (Kansas), 12 September 1983, p. 12. The 0.7 ounces of fish per capita daily can be calculated from U.S. Department of Agriculture, *Agricultural Statistics* (Washington, D.C.: U.S. Government Printing Office, 1981), p. 552. The ten million acres required to feed the eight million horses is obtained by multiplying forty-five million acres by 8/25 by 2/3, with the assumption that a recreational horse eats two-thirds of what a draft horse eats.

25. The nineteen million acres of feed needed for eight million working horses and mules in Table 7 are obtained by multiplying forty-five million acres by 8/25 by 4/3. The 242 million acres needed for 10.2 ounces of meat are obtained by multiplying 142 million acres by 300/234 by 4/3. The acreages for dairy cattle and direct human consumption are calculated in a similar manner.

26. For the recommended four ounces of meat, see Mary P. Clarke, "Having Your Protein and Lower Costs Too," rev. Cooperative Extension Service brochure L-413 (Manhattan: Kansas State University, 1979). The current average daily diet of 3,300 calories and 100 grams of protein is obtained from Louise Page

and Berta Friend, "The Changing United States Diet," *Bioscience* 28, no. 3 (1978): pp. 192–197. The 2,860 calories and 72 grams of protein can be calculated from U.S. Department of Agriculture, *Agricultural Statistics* (Washington, D.C.: U.S. Government Printing Office, 1977), p. 562. The world per capita diet in 1971 of 2,270 calories and 65 grams of protein was obtained from C.R.W. Spedding, J.M. Walsingham, and A.M. Hoxey, *Biological Efficiency in Agriculture* (New York: Academic Press, 1981), table 10.1, p. 121.

27. The proportion that each kind of livestock and poultry contributes to our meat diet (excluding fish) can be calculated from USDA, *Agricultural Statistics*, in note 24, p. 257, 564.

28. The 1920s level of 3,460 calories and 95 grams of protein is from Page and Friend, "The Changing United States Diet," p. 194.

29. The seven ounces of meat can be calculated from Page and Friend, "The Changing United States Diet," p. 192.

30. Sylvan H. Wittwer, "Assessment of Technology in Food Production" in *Renewable Resource Management for Forestry and Agriculture*, eds. Bethel and Massengale, p. 46.

NOTES TO CHAPTER ELEVEN

1. Cynthia Hewitt de Alcantara, *Modernizing Mexican Agriculture: Socio-Economic Implications of Technological Change, 1940–1970*. (Geneva: United Nations Research Institute on Rural Development, 1976); Ernest Feder, "La nueva penetracion en la agricultura de los paises subdesarrollados por los paises industriales y sus empresas multinacionales," *Revista del Mexico Agrario* 9, no. 3 (May/June 1976): 101–141; Sergio Reyes Osorio et al., *Estructura Agraria y Desarrollo Agricola en Mexico* (Mexico City: Fondo de Cultura Economica, 1974); Peter Baird and Ed McCaughan, *Beyond the Border* (New York: North American Congress on Latin America, 1979); Phillip Russell, *Mexico in Transition* (Austin, Tex.: Colorado River Press, 1977); for a typical example of the dismissal of the charges against American research in Mexico, see Vance Bourjaily, "One of the Green Revolution Boys," *Atlantic Monthly* 231, no. 2 (February 1973): 66–76.

2. Paul Lamartine Yates, *Mexico's Agricultural Dilemma* (Tucson: University of Arizona Press, 1981), p. 10.

3. Bourjaily, "One of the Green Revolution Boys," pp. 66–76. The original authors of the Rockefeller program evinced more ambiguity on this point than Borlaug, the program's chief scientist. They feared the results of excessive land reform, creating too many plots not large enough to support families. See E.C. Stakman, Richard Bradfield, and Paul C. Mangelsdorf, *Campaigns Against Hunger* (Cambridge, Mass: Belknap Press, 1967), p. 22.

4. Edmund K. Oasa and Bruce H. Jennings, "Science and Authority in International Agricultural Research," *Bulletin of Concerned Asian Scholars* 14 (October/December 1982): 32–33. An earlier version of this article appeared in *El Trimestre Economico*, no. 196, 1982. The authors provide powerful insight

into the Rockefeller Foundation's work in both Mexico at CIMMYT and in the Philippines at the International Rice Research Institute.

5. Oasa and Jennings, "Science and Authority," p. 33.
6. Oasa and Jennings, "Science and Authority," pp. 35–36.
7. Oasa and Jennings, "Science and Authority," p. 39.
8. Yates, *Mexico's Agricultural Dilemma*, p. 11.
9. Yates, *Mexico's Agricultural Dilemma*, p. 1.
10. Yates, *Mexico's Agricultural Dilemma*, p. 11.
11. David Ronfeldt, *Atencingo: The Politics of Agrarian Struggle in a Mexican Ejido* (Stanford, Calif.: Stanford University Press, 1973), p. 221.
12. Keith Griffin, *The Green Revolution: An Economic Analysis* (Geneva: United Nations Research in Social Development, 1972). The suggestions with respect to Poland come from Carol Naggengast of the anthropology department of the University of California, Irvine, who will shortly finish her Ph.D. dissertation on Polish agricultural development.
13. Bourjaily, "One of the Green Revolution Boys," p. 75.

NOTES TO CHAPTER THIRTEEN

1. Gene C. Wilken, "Some Aspects of Resource Management by Traditional Farmers," in *Small Farm Agricultural Development Problems*, eds. H.H. Biggs and R.L. Tinnermeier (Fort Collins: Colorado State University Press, 1974), pp. 47–59; Stephen R. Gliessman and Moises Amador Alarcon, "Ecological Aspects of Production in Traditional Agroecosystems in the Humid Lowland Tropics of Mexico," in *Tropical Ecology and Development*, ed. José I. Furtado (Kuala Lumpur, Malaysia: International Society for Tropical Ecology, 1980), pp. 601–608; and Gary Klee, *World Systems of Traditional Resource Management* (New York: Halstead, 1980).

2. Stephen R. Gliessman, Roberto Garcia Espinosa, and Moises Amador Alarcon, "The Ecological Basis for the Application of Traditional Agricultural Technology in the Management of Tropical Agro-Ecosystems," *Agro-Ecosystems* 7, no. 3 (1981): 173–185; John Lambert, "The Ecological Consequences of Ancient Maya Agricultural Practices in Belize, Central America," presented at the symposium on Prehistoric Intensive Agriculture in the Tropics, Australian National University, Canberra, August 1981.

3. Orie Loucks, "Emergence of Research on Agro-Ecosystems," *Annual Review of Ecology and Systematics* 8 (1977): 173–192; Efraim Hernandez Xolocotzl, "Agroecosistemas de México: contribuciones a la enseñanza, investigación, y divulgación agrícola" (Colegio de Postgraduados, Chapingo, Mexico, 1977); C.R.W. Spedding, *The Biology of Agricultural Systems* (London: Academic Press, 1975); Robert Hart, "Agroecosistemas: conceptos básicos," (Centro Agrinomico Tropical de Investigación y Enseñanza, Turrialba, Costa Rica, 1979); and Gordon Conway, "What Is an Agroecosystem and Why Is It Worthy of Study?," delivered at the Workshop on Human/Agroecosystem Interactions, Program on Environmental Science and Man-

agement/East West Environmental and Policy Institute, Los Banos College, Laguna, Philippines.

4. Peter Harrison and Billie Lee Turner III, eds., *Pre-Hispanic Maya Agriculture* (Albuquerque: University of New Mexico Press, 1978).

5. Brian Trenbath, "Biomass Productivity in Mixtures," *Advances in Agronomy* 26 (1974): 177–210.

6. Moises Amador Alarcon, "Comportamiento de tres especies (Maiz, Frijol, Calabaza) en policultivos en la Chontalpa, Tabasco, México" (Tesis Profesional, Cardenas, Tabasco, Mexico, 1980).

7. Douglas Boucher and Judith Espinosa, "Cropping System and Growth and Nodulation Responses of Beans to Nitrogen in Tabasco, Mexico," *Tropical Agriculture* 59, no. 4 (1982): 279–282; and Douglas Boucher, "Nodulation of Beans in Polyculture: The Effect of Distance Between Plants of Bean and Corn," *Agricultura Tropical* (Colegio Superior de Agricultura Tropical) 1 (1979): 26–283 (in Spanish).

8. Stephen Risch, "The Population Dynamics of Several Herbivorous Beetles in a Tropical Agro-Ecosystem: The Effect of Intercropping Corn, Beans, and Squash in Costa Rica," *Journal of Applied Ecology* 17 (1980): 593–612.

9. Deborah Letourneau, "Population Dynamics of Insect Pests and Natural Control in Traditional Agroecosystems in Tropical Mexico" (Ph.D. diss., University of California, Berkeley, 1983); and Miguel Angel Altieri, "Weeds May Augment Biological Control of Insects," *California Agriculture* 35, nos. 5/6 (May/June 1981): 22–24.

10. Stephen R. Gliessman and Moises Amador Alarcon, "A Traditional Pest Management Strategy Using Alarm Pheromones," for *Biotropica*, manuscript in preparation.

11. See reviews on the subject of pheromones in William L. Brown, Jr., Thomas Eisner, and Robert H. Whittaker, "Allomones and Kairomones: Transspecific Chemical Messengers," *Bioscience* 20, no. 1 (1970): 319–322; Robert H. Whittaker and Paul P. Feeny, "Allelochemic Chemical Interactions Between Species," *Science* 171, no. 3973 (1971): 757–770.

12. Kathryn Yih et al., "Weed Control and Polycultural Advantage," *Biological Agriculture and Horticulture* (1983), in press.

13. Juan Carlos Chacon and Stephen R. Gliessman, "Use of the 'Non-Weed' Concept in Traditional Agroecosystems of South-Eastern Mexico," *Agro-Ecosystems* 8, no. 1 (1982): 1–11.

14. John Ewel et al., "Leaf Area, Light Transmission, Roots and Leaf Damage in Nine Tropical Plant Communities," *Agro-Ecosystems* 7, no. 4 (1982): 305–326; Stephen R. Gliessman, "Allelopathic Interactions in Crop/Weed Mixtures: Applications for Weed Management," *Journal of Chemical Ecology* 9, no. 8 (1983): 990–991.

15. John Vandermeer, "The Interference Production Principle: An Ecological Theory for Agriculture," *Bioscience* 31, no. 5 (1981): 361–364.

16. Stephen R. Gliessman and Miguel Angel Altieri, "Advantages of Polyculture Cropping," *California Agriculture* 36, no. 7 (July 1982): 14–17.

17. Gliessman, "Allelopathic Interactions."
18. Peter Price, "Colonization of Crops by Arthropods: Non-Equilibrium Communities in Soybean Fields," *Environmental Entomology* 5, no. 4 (1976): 605–611.

NOTES TO CHAPTER FOURTEEN

1. R.D. Hodges, "Agriculture and Horticulture: The Need for a More Biological Approach," *Biological Agriculture and Horticulture* 1, no. 1 (1982): 1–13.
2. Regarding water and energy, see K.E. Foster and N.G. Wright, "Constraints to Arizona Agriculture," *Journal of Arid Environments* 3 (1980): 85–94. On Arizona's energy costs, see U.S. Department of Agriculture and the Federal Energy Commission, *U.S. Agriculture: 1974 Data Base* (Washington, D.C.: U.S. Government Printing Office, 1976). On humid-adapted plants in arid lands, see David S. Pimentel et al., "Water Resources in Food and Energy Production," *Science* 32, no. 11 (1982): 861–866. And on specialty crops, see A.A. Theisen, E.G. Knox, and F.L. Mann, *Feasibility of Introducing Food Crops Better Adapted to Environmental Stress* (Washington, D.C.: National Science Foundation and U.S. Government Printing Office, 1978).
3. Gary P. Nabhan, "Seeds of Prehistory," *Garden* 4, no. 3 (1980): 8–12.
4. George W. Cox and Michael D. Atkins, "Agricultural Ecology," *Bulletin of the Ecological Society of America* 56, no. 3 (1975): 2–6.
5. Immanuel Noy-Meir, "Desert Ecosystems I: Environment and Producers," *Annual Review of Ecology and Systematics* 4 (1973): 25–52.
6. Daniel H. Janzen, "Tropical Agroecosystems," *Science* 182 (1973): 1212–1219.
7. M. Evenari, L. Shanan, and N. Tadmor, *The Negev—The Challenge of a Desert* (Cambridge, Mass.: Harvard University Press, 1971).
8. J.A. Ludwig and Stan D. Smith, "Comparative Primary Production of Chihuahuan Desert Ecosystems," *Bulletin of the New Mexico Academy of Sciences* 18 (1978): 8–11.
9. K.E. Foster, R.L. Rawles, and M.M. Karpiseak, "Biomass Potential in Arizona," *Desert Plants* 2, no. 3 (1980): 197–200.
10. On nitrogen-fixing plants, see, for example, P. Felker and P.R. Clark, "Nitrogen Fixation (Acetylene Reduction) and Cross Inoculation in 12 *Prosopis* (Mesquite) Species," *Plant and Soil* 57 (1980): 177–186. On organics in alluvial fans, see Gary P. Nabhan, "Papago Indian Fields: Arid Lands Ethnobotany and Agricultural Ecology" (Ph.D. diss., University of Arizona, 1983). On nitrogen contents of livestock manure, see Soil and Health Society, *You Can Have Healthy Soil* (Emmaus, Pa.: Soil and Health Society, 1981).
11. Forrest Shreve, "Vegetation of the Sonoran Desert," in *Vegetation and Flora of the Sonoran Desert*, eds. Forrest Shreve and Ira W. Wiggins (Stanford, Calif.: Stanford University Press, 1964), pp. 1–186; and Nabhan, "Papago Indian Fields."
12. L.M. Thompson and F.R. Troeh, *Soils and Soil Fertility* (New York: McGraw-Hill, 1978).

13. W.A. Cannon, "The Root Habits of Desert Plants," Carnegie Institute of Washington, Publication 131 (1911): 1–96.

14. Otto T. Solbrig, "The Strategies and Community Patterns of Desert Plants," in *Convergent Evolution in Warm Deserts*, eds. Gordon Orians and Otto T. Solbrig (Stroudsburg, Pa.: Dowden, Hutchinson and Ross, 1978), pp. 67–106.

15. Neil F. Hadley and Stan R. Szarek, "Productivity of Desert Ecosystems," *Bioscience* 31, no. 10 (1981): 747–750.

16. R.W. Snaydon, "Plant Demography in Agricultural Systems," in *Evolution in Plant Populations*, ed. Otto T. Solbrig (Berkeley: University of California Press, 1980), pp. 131–160.

17. Richard S. Felger, "Ancient Crops for the Twenty-first Century," in *New Agricultural Crops*, ed. Gary A. Ritchie (Boulder, Colo.: Westview Press, 1979), pp. 5–20; and Richard S. Felger and Gary P. Nabhan, "Agroecosystem Diversity: A Model from the Sonoran Desert," in *Social and Technological Management in Dry Lands*, ed. Nancie L. Gonzales (Boulder, Colo.: Westview Press, 1978), pp. 129–149.

18. On mesquite, see D.O. Cornejo et al., "Utilization of Mesquite in the Sonoran Desert: Past and Future," in *Mesquite Utilization—1982*, ed. Harry W. Parker (Lubbock: College of Agricultural Sciences, Texas Tech University, 1982), pp. 1–20. Regarding agave, see H.S. Gentry, *Agaves of Continental North America* (Tucson: University of Arizona Press, 1982). Regarding devil's claw, see James Berry et al., "Domesticated *Proboscidea Parviflora*: A Potential Oilseed Crop for Arid Lands," *Journal of Arid Environments* 4, no. 2 (1981): 147–160. On yields from polyculture, see Miguel A. Altieri, Deborah K. Letourneau, and James A. Davis, "Developing Sustainable Agroecosystems," *Bioscience* 33, no. 1 (1983): 45–49.

19. For an overview of tepary beans, with notes on nitrogen, see Gary P. Nabhan, ed., "The Desert Tepary as a Food Resource—A Journal Symposium," *Desert Plants* 5, no. 1 (1983): 1–64.

20. Paul D. Hurd, Jr., E. Gordon Linsley, and Thomas W. Whitaker, "Squash and Gourd Bees (*Peponapis, Xenoglossa*) and the Origin of Cultivated *Cucurbita*," *Evolution* 25, no. 1 (1971): 218.

21. Paul D. Hurd, Jr., and E. Gordon Linsley, "Pollination of the Unicorn Plant (*Martyniaceae*) by an Oligolectic Corolla-cutting Bee (*Hymenoptera: Apoideae*)," *Journal of the Kansas Entomological Society* 36, no. 4 (1963).

NOTES TO CHAPTER FIFTEEN

1. For buffalo grass yields, see Robert M. Ahring, *The Management of Buffalo Grass for Seed Production in Oklahoma*, Technical Bulletin, T-109 (Stillwater: Oklahoma Agriculture Experiment Station, 1964). For Alta fescue and sand dropseed, see H. Ray Brown, "Growth and Seed Yields of Native Prairie Plants in Various Habitats of the Mixed Prairie," *Transactions of the Kansas Academy of Science* 46 (1943): 87–99.

2. For Illinois bundleflower, see U.S. Department of Agriculture, *Annual Report*

(Knox City, Tex.: Plant Materials Center, Soil Conservation Service, 1978). For cicer milk vetch, see U.S. Department of Agriculture, *Technical Note 12/1/66* (Bridger, Mont.: Plant Materials Center, Soil Conservation Service, 1966). For sanfoin, see Ashley Thornberg, *Grass and Legume Seed Production*, Bulletin no. 333 (Bozeman: Montana Agricultural Experiment Station, 1971).

3. F.R. Earle and Quentin Jones, "Analyses of Seed Samples from 113 Plant Families," *Economic Botany* 16, no. 4 (1962): 221–250; and Quentin Jones and F.R. Earle, "Chemical Analyses of Seed II: Oil and Protein Content of 759 Species," *Economic Botany* 20, no. 2 (1966): 127–155.

4. See D.C.L. Kass, "Polyculture Cropping Systems: Review and Analysis," *Agriculture Bulletin* 32, no. 55 (1978).

5. Jack R. Harlan, *Theory and Dynamics of Grassland Agriculture* (New York: D. Van Nostrand, 1956), p. 37.

6. Wes Jackson and Marty Bender, "Horses or Horsepower?," *Soft Energy Notes* 5, no. 3 (1982): 70–87.

7. U.S. Department of Agriculture, *Soil and Water Resource Conservation Act, 1980 Appraisal* (Washington, D.C.: U.S. Department of Agriculture, 1980), pt. 1.

8. U.S. Department of Agriculture and Federal Energy Administration, *Energy and Agriculture: 1974 Data Base* (Washington, D.C.: U.S. Department of Agriculture, 1976), vol. 2.

9. USDA, *Energy and Agriculture*, vol. 1, p. 14.

10. Wes Jackson and Marty Bender, "New Roots for American Agriculture" (Contract paper for the U.S. Office of Technology Assessment, 1980).

11. USDA, *Energy and Agriculture*, vol. 1.

12. USDA, *Energy and Agriculture*, vol. 1.

13. U.S. Water Resources Council, *The Nation's Water Resources: 1975–2000, Second National Water Assessment* (Washington, D.C.: U.S. Government Printing Office, 1974), vol. 1, p. 14; Alexander Zaporozec, "Changing Patterns in Ground-Water Use in the United States," *Ground Water* 17, no. 2 (1979): 200; Gordon Sloggett, "Energy Used for Pumping Irrigation Water in the United States, 1974," in *Agriculture and Energy: Proceedings of a Conference*, ed. William Lockeretz (New York: Academic Press, 1977), pp. 115, 117; and USDA, *Energy and Agriculture*, vol. 1, p. 12.

NOTES TO CHAPTER SEVENTEEN

1. Arnold M. Schultz, "The Ecosystem as a Conceptual Tool in the Management of Natural Resources," in *Natural Resources: Quality and Quantity*, eds. S.V. Ciriacy Wantrup and James S. Parsons (Berkeley: University of California Press, 1967).

2. A.G. Tansley, "The Use and Abuse of Vegetational Concepts and Terms," *Ecology* 16 (1935): 284–307.

3. Cited in Schultz, "The Ecosystem as a Conceptual Tool," p. 141.

4. J.S. Rowe, "The Level of Integration Concept and Ecology," *Ecology* 42 (1961): 420–427.
5. J.K. Feibleman, "Theory of Integrative Levels," *British Journal of the Philosophy of Science* 5 (1954): 59–66.
6. In a comparison between old order Amish and corn belt farmers in northern Indiana, wildlife habitat interspersion was found to be three times greater on Amish farms. Smaller fields and three times as much edge are responsible for this much extra wildness. This is based on work by Mark Biggs, "Conservation Farmland Management—The Amish Family Farm Versus Modern Corn Belt Agriculture" (M.S. thesis, School of Forest Resources, Pennsylvania State University, May 1981).
7. P.W. Anderson, "More Is Different," *Science* 177 (1972): 393–396.
8. E.P. Odum, *Fundamentals of Ecology* (Philadephia: W.B. Saunders, 1971), p. 5.
9. F.A. Bazzaz and J.A.D. Parrish, "Organization of Grassland Communities," in *Grasses and Grasslands: Systematics and Ecology*, eds. James Estes et al. (Norman: University of Oklahoma Press, 1982).
10. E.O. Wiley and Daniel R. Brooks, "Victims of History—A Nonequilibrium Approach to Evolution," *Systematic Zoology* 31, no. 1 (1982): 1–24.
11. R.M. May, *Stability and Complexity in Model Ecosystems* (Princeton, N.J.: Princeton University Press, 1973).
12. Wendell Berry, "An Agricultural Journey in Peru," in *The Gift of Good Land* (San Francisco: North Point Press, 1981).
13. Wendell Berry, "People, Land and Community," in *Standing by Words* (San Francisco: North Point Press, 1983), p. 76. Emphasis added.

Contributors

EDITORS

Wes Jackson is founder and codirector of The Land Institute, on the banks of the Smoky Hill River, near Salina, Kansas. He is on the boards of directors of several environmental organizations and is author of *New Roots for Agriculture* (1980) and editor of *Man and the Environment* (1971).

Wendell Berry is a lifelong Kentuckian who farms in Henry County. He has taught for a number of years at the University of Kentucky. His novels include *A Place on Earth* (1983); among his books of poetry are *A Part* (1980) and *The Wheel* (1982). His writing on farming includes *The Unsettling of America* (1977) and *The Gift of Good Land* (1981).

Bruce Colman, a third-generation Californian, is an editor and critic who lives in Berkeley. For over a decade, he worked for Friends of the Earth; he now does publishing consulting and freelance work.

AUTHORS

Marty Bender is a research associate at The Land Institute. A physicist and chemist by training, he is coauthor with Wes Jackson of several articles in *Soft Energy Notes* and *The Journal of Soil and Water Conservation*. He was born in Dayton, Ohio.

Jennie E. Gerard was born in Kentucky and now lives in the San Franciso Bay Area. She is Vice President of The Trust for Public Land.

Stephen Gliessman earned his doctorate in plant ecology at the University of California, Santa Barbara, and is now director of the Agroecology Program at U.C. Santa Cruz.

Dana Jackson is codirector of The Land Institute and editor of *The Land Report*.

Hans Jenny is emeritus professor of soil science at the University of California and former president of the Soil Science Society of America. He is author of *The Soil Resource: Origin and Behavior* (1980).

Sharon Johnson holds a master's degree in geography and human environmental studies. She is a writer for The Trust for Public Land.

Gene Logsdon lives and farms in Ohio; he is author of *Two Acre Eden* (1978) and *A Grove of Trees to Live In* (1981). He is a regular contributor to Rodale publications, *Country Journal*, and *Ohio* magazine.

Amory B. and *L. Hunter Lovins* are codirectors of the Rocky Mountain Institute, near Snowmass, Colorado, and policy consultants to Friends of the Earth. Together, they wrote *Energy/War* (1980). Mr. Lovins is author of *Soft Energy Paths* (1977). Mrs. Lovins, a lawyer, was for six years assistant director of the "Tree People" conservation group in California.

Gary Paul Nabhan, author of *The Desert Smells Like Rain* (1982), is president of Native Seeds/SEARCH, a nonprofit conservation organization. He teaches at the arid lands studies department of the University of Arizona.

Gary Snyder, a leading American student of Buddhism, lives in the Sierra Nevada foothills. His books of poetry include *Turtle Island*, which won the 1974 Pulitzer Prize for poetry, and *Axe Handles* (1983).

Marty Strange is codirector of The Center for Rural Affairs, in Walthill, Nebraska.

John Todd is founder of The New Alchemy Institute, the Company of Stewards, Inc., and Ocean Arks International. With his wife, Nancy, he wrote *Tomorrow Is Our Permanent Address* (1980).

Donald Worster comes from a rural Kansas family. He teaches American studies at the University of Hawaii, and his books include *Dust Bowl* (1979) and *Nature's Economy* (1977). *Dust Bowl* won Columbia University's Bancroft Prize for distinguished work in history.

Angus Wright was born in Salina, Kansas. He is director of the Environmental Studies Center at California State University, Sacramento.

Design by David Bullen
Typeset in Mergenthaler Sabon
By G & S Typesetters
Printed by Maple-Vail
on acid-free paper